The Once and Future
MOON

The Once and Future
MOON

☾

Paul D. Spudis

Smithsonian Institution Press
Washington and London

Copy Editor: D. Teddy Diggs
Production Editor: Duke Johns
Designer: Linda McKnight

Library of Congress Cataloging-in-Publication Data
Spudis, Paul D.
    The once and future moon / Paul D. Spudis.
      p. cm. — (Smithsonian library of the solar system)
    Includes bibliographical references and index.
    ISBN 1-56098-634-4 (alk. paper)
    1. Moon. I. Title. II. Series.
    Q581.S686   1996
    523.3—dc20   95-51343

British Library Cataloguing-in-Publication Data is available

Manufactured in the United States of America
03 02 01 00 99 98 97 96 5 4 3 2 1

⊚ The paper used in this publication meets the minimum requirements of the American National Standard for Information Sciences—Permanence of Paper for Printed Library Materials ANSI Z39.48-1984.

To Anne
Loving wife, insightful critic, and best friend

Contents

Preface

The number of books written about the Moon* over the past 30 years could easily fill a small library. They cover a variety of topics, including craters, the surface, rocks, chemistry, minerals, the interior, and space travel to and from our nearest planetary neighbor. In the past few years several books have been published that describe the U.S.-Soviet "Moon race" of the 1960s and attempt to re-create the atmosphere of that distant time, when the two global superpowers fought for "control of the heavens." Other books are technical in nature and cogently summarize our understanding of the geological story of the Moon or describe how we can establish a base on the Moon and detail some of the activities we might undertake there.

Given all of these previous efforts, why write yet another book about the Moon? Although there are several excellent books about the Moon written for the intelligent adult, they were mostly published several years to over a decade ago. Progress marches on in lunar science, and now, a quarter-century after the Apollo program, is an appropriate time to summarize our current understanding of the Moon's history and evolution. What's more, we have just revisited the Moon for the first time since Apollo with the Department of Defense Clementine mission in 1994. This spacecraft mapped the composition and shape of the entire Moon and has provided us with some startling new insights into lunar processes and history. I was fortunate to have

---

*By "the Moon," I mean Luna, the natural satellite of the Earth. This object is a complex planetary body with its own history of geological evolution and was humanity's first stepping stone into the universe. It deserves the dignity of capitalization. The generic term "moon" refers to any natural satellite of any planet.

been a member of the Science Team for this mission, and I have incorporated many of the new insights that we obtained from Clementine into the text. I believe that Clementine will cause us to look at the Moon and its history in a new light after all of the data have been digested in a few years. In this book I have tried to capture some hints of how significant such a revision of our understanding will be.

Another development in the field of lunar studies has been the movement for a return to the Moon, including the undertaking of both robotic and human missions. This campaign began in the early 1980s, in particular at the Lunar Base Symposium held in 1984. I have been involved in this effort since then, including doing a year-long stint in Washington in 1991 as a member of the Synthesis Group, a White House committee headed by the astronaut Tom Stafford. This panel of people from a wide variety of defense and civilian aerospace fields developed various plans designed to implement the ill-fated Space Exploration Initiative, the proposal by President George Bush to establish a base or outpost on the Moon and to conduct a human mission to Mars. Having observed the fate of this Initiative and other attempts to resume lunar exploration, I believe that I know why such efforts failed and why we are not on the Moon today. No one has told this story before, a story that I believe has implications both for the current state of the National Aeronautics and Space Administration (NASA) and of the space program and for the future of our nation in space.

This book aims to tell the story of our first visits to the Moon, what we learned about it as a result, and why and how we might someday return there. My title is chosen both to pay homage to T. H. White's wonderful book on King Arthur and to emphasize my belief that the Moon holds an important place in our space future. I have found my calling in the fascinating and challenging study of lunar history and processes, but there are many other aspects to the Moon's importance. It is not only a place of great wonder and beauty but also a strategic and valuable planetary object. I hope this book will kindle both an increased understanding of and a new interest in the Moon.

Several people have read various versions of this manuscript, and their advice has improved the text in many ways. My wife,

Anne, is both a constructive critic and a careful reader; I could not have written this book without her advice, encouragement, and support. I have also received the benefit of positive criticism from David Black, Nancy Ann Budden, Bret Drake, Jeffrey Gillis, John Gruener, Graham Ryder, Buck Sharpton, Karen Stockstill, and S. Ross Taylor, and I thank them all for helping me to avoid the more obvious errors of fact and awkward expression. Don Wilhelms, my mentor and colleague in the lunar business, critically read an early version of the manuscript and made many helpful suggestions. I also thank Grant Heiken, Ted Maxwell, and Jeff Taylor for review of and comment on the manuscript. Several friends and colleagues have provided figures for this book. I thank Don Campbell (10.1), Dave Criswell (9.7), Don Davis (Plate 6), Bill and Sally Fletcher (Plate 1), Ron Greeley (5.1), B. Ray Hawke (5.11), Stewart Nozette (11.1), Pat Rawlings (9.4, 9.5, 9.6, 9.8. 9.9, 10.2, 10.4, 10.6, 10.8), Graham Ryder (6.9, Plate 4), and John Wood (4.3). The opinions expressed in this book are entirely my own and are not necessarily those of my reviewers or colleagues or the institution for which I work.

 Chapter 1

To Study the Moon

The Moon has fascinated and puzzled humanity for millennia, but the history of its scientific study is actually rather recent. It was not until 1610, when Galileo Galilei turned his "spyglass" toward the Moon, that the systematic description and study of its surface features was even possible. Some of the earliest investigations were quite insightful, correctly deducing such aspects as the relative age of its surface features and the origin of its craters by impact. Here I touch on some of the highlights of this long and fascinating history, which led to our first understanding of another world in the solar system.

The Watchers and the Mappers

People have gazed on the Moon for centuries, watching it wax (appear to grow larger) and wane (appear to shrink) in the sky with the passage of every *month* (a word itself derived from the word *moon*). Even to the naked eye, the Moon shows a contrasting pattern of light and dark patches. This pattern, fancifully interpreted by different cultures at various times, has represented people, rabbits, frogs, crickets, and a host of other mythical lunar inhabitants. The Moon has been worshiped as a deity, typically as a goddess, in almost all cultures. The occasional (and terrifying) eclipses of the Moon, inexplicable and unpredictable to primitive peoples, befit the image of a deity displeased with her followers and their lack of devotion. This personification of the Moon as female is probably related to its regular, monthly phases, so nearly coincident in duration with a woman's cycle.

To the Greek philosophers, it was clear that the Moon is a

heavenly body that orbits Earth, is a sphere, and shines in the night sky by reflected sunlight. From the shape of the shadow of Earth cast on the Moon during lunar eclipses, the Greek astronomer Aristarchus (third century B.C.) estimated that the Moon was about 60 Earth radii from Earth, an excellent determination of its true distance, which varies between about 55 and 63 Earth radii (354,000 to 404,000 km). Some Greek philosophers believed that the Moon was a world much like our own, merely distant from us. Plutarch (first century A.D.) even went so far as to suggest that the Moon was inhabited by people! The Greeks also apparently believed that the dark areas were seas and the bright regions were land. A memory of this concept remains in lunar science today in the Latin names that we give to these areas: *maria* (seas) for the dark regions and *terrae* (lands) for the bright, rugged highlands (Plate 1).

In the cosmology of Ptolemy (second century A.D.), the great astronomer of the ancient world, the Moon was Earth's nearest neighbor in space, orbiting Earth, as did its daytime neighbor, the Sun. A revolution in astronomical thought, initiated by Copernicus in the 16th century, consigned Earth to a subsidiary role in the solar system by having it orbit the Sun, but this model kept the Moon firmly anchored to Earth. Galileo is commonly credited with the first scientific description of the Moon from telescopic observation (Fig. 1.1). He published his observations in 1610 in the book *Sidereus Nuncius* (Celestial Messenger). Galileo described a rough and mountainous surface, quite different from the expectations of a perfect, smooth celestial body. He noted that the light and dark regions of the Moon were areas of rough, hilly topography and smoother plains, respectively. Galileo was particularly intrigued with the revelation of high lunar mountains, and his detailed description of a large crater in the central highlands (probably Albategnius, see Fig. 1.1) prefigured 350 years of controversy and debate about the origin of these strange holes on the Moon.

Having been shown the way by Galileo, the astronomers of the 17th century spent much effort mapping and cataloging every lunar surface feature, seen at ever greater detail by increasingly powerful telescopes. The Flemish astronomer Langrenus published a map in 1645 that gave names to the surface features of the

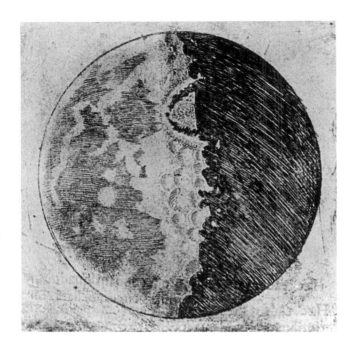

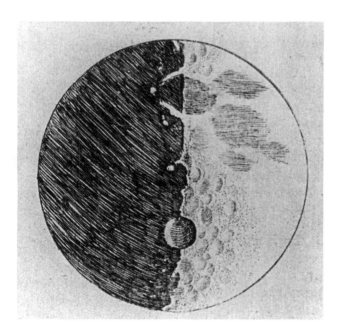

Figure 1.1. Two views of the Moon, drawn by Galileo in 1610. In the top view, Mare Serenitatis is clearly visible at top center. In the bottom drawing, the large crater is probably Albategnius; its prominence is somewhat exaggerated.

Figure 1.2. Map of the Moon, drawn in 1647 by the Polish astronomer Höwelcke (Latin name: Hevelius), who showed the libration zones, the effect that permits us to see partly around to the far side.

Moon (mostly to its craters). Because Langrenus's patron was the king of Spain, his scheme of names honored the king and the rather extended royal family; however, the only one of his names that survives today is attached to the crater that Langrenus named for himself. The map drawn by Rheita (1645) correctly depicts the bright ray systems of the fresh craters Tycho and Copernicus. An effort by Hevelius (1647) included the libration zones of the Moon (Fig. 1.2), the edges that wobble into view and occasionally allow us to see over 50 percent of the Moon's surface. By 1651 the astronomers Giambattista Riccioli and Francesco Grimaldi had published a map (Fig. 1.3) that established the scheme of nomenclature we use today: craters are named for famous scientists (e.g., Copernicus, Archimedes), and the dark regions (maria) are given classical Latin names denoting the

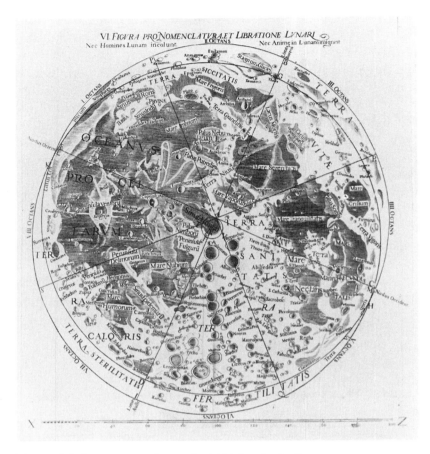

Figure 1.3. Lunar map drawn in 1651 by the Italian astronomer Grimaldi, showing the nomenclature scheme that was devised by Riccioli and that is still used today. Craters are named for famous scientists and mathematicians, and the maria are given names denoting states of mind or the weather.

weather (e.g., *Mare Imbrium*, Sea of Rains) or states of mind (e.g., *Mare Tranquillitatis*, Sea of Tranquillity).

Mapping the Moon continued sporadically throughout the intervening years, striving to portray the surface in ever increasing detail. The German astronomers Wilhelm Beer and J. H. von Mädler produced a map (1834) of the Moon that codified the naming scheme for surface features while pushing the limits of telescopic observation. The subsequent 1878 map by J. F. Schmidt was over 100 cm (40 inches) in diameter and showed features as small as a few hundred meters across. The tendency of astronomers in this era was to attempt to see smaller and smaller features, the idea being that the secrets of lunar history were in the details. In fact much of this was misplaced effort. Today we understand that it is the broad-scale features of the Moon that tell its story most eloquently. Mapping the Moon with telescopes on Earth culminated to near perfection with the publication (1960, 1967) of the magnificent *Photographic Lunar Atlas* (jointly published by the U.S. Air Force and the University of Arizona), a collection of the very best images obtained from all of Earth's observatories. This atlas was the last to be prepared largely before the advent of the space age, after which we could examine the Moon's surface up close. The *Photographic Lunar Atlas* is still useful today because it depicts the near side of the Moon under a variety of lighting conditions (useful for studies of surface shape and process), something rarely attained during spacecraft exploration because of the short duration of missions. It is a rare publication and is highly prized by the lucky few who own copies.

Lunacy: Then and Now

The concept that the Moon is responsible for a variety of ill and strange effects on people is very old and is closely wrapped up in mythology. For some reason the Moon, particularly a full moon, is responsible for vampires, werewolves, madness, ritual murder—you name it. This association is probably related to a primeval human fear of the dark and of night, when the unseen surrounds us and when bad things happen. A full moon is bright enough to allow people to roam about at night, certainly an "unnatural" condition. This long-popular association of the Moon with strangeness or madness continued throughout the era of modern scientific study and, indeed, continues to this day, as evidenced by the myth that crime rates and strange behavior increase during periods of a full moon.

Outside of the gripping science fiction of Jules Verne and H. G. Wells, one of the most memorable Moon myths was created by a journalist (naturally) who published a series of articles in 1835 in the *New York Sun* newspaper. These articles were based on alleged reports from the astronomer John Herschel (son of William Herschel, the discoverer of the planet Uranus), who had begun a series of observations from South Africa (at the time, most of the southern sky was unknown to the scientific community). The reporter, Richard Locke, claimed that Herschel had devised the most powerful telescope ever made and was seeing astonishing things: strange plants, animals, and flying, batlike creatures that appeared to be intelligent—all inhabiting the Moon! Needless to say, this astonishing discovery caused sales of the *Sun* to skyrocket, the intended consequence of the article series. Herschel was conveniently unavailable for confirmation of the discoveries. After a month, the hoax was finally exposed by another reporter. When a group of scientists demanded to see the original reports from Herschel, Locke was forced to admit that he had made up the whole thing. This episode was a milestone in the history of skeptical debunking.

Lunar scientists were not (and are not) immune to crazy discoveries. The German astronomer Franz von Paula Gruithuisen was a competent and careful observer. He was the first to suggest that craters on the Moon were formed by the impact of meteorites. However, in 1848 Gruithuisen announced that he had discovered walled cities on the Moon! His lunar metropolis was one of the earliest examples of otherwise careful workers trying a little too hard to make new and significant discoveries. A persistent and wrong idea (not disproved until the space age) was the assertion that the crater Linné in Mare Serenitatis appeared and disappeared on an irregular basis. We now believe that this very bright, fresh crater, which is just at the limits of visual perception in Earth-based telescopes, is indeed a constant feature. Its unusual disappearances and reappearances are probably an optical illusion caused by poor astronomical "seeing."

Even today something about the Moon continues to attract the fringe mentality. Many books and newspapers advocate the existence of secret UFO bases at the lunar north pole, strange engineering projects on the far side of the Moon, artifacts protruding from the Moon hundreds of meters into space, and human skeletons identified in "secret" NASA photos of the Moon. It is claimed that extensive cover-ups and giant conspiracies by the U.S. government prevent the American people from discovering the real truth about aliens on the Moon (an amusing concept to anyone who has ever actually worked for the federal government and has tried to get even the simplest things done). According to these accounts, the aliens on the Moon (1) want to be our friends, (2) are our enemies, (3) are completely indifferent to our existence, and/or (4) want our women (pick one or more). Many of these moronic morsels are still in print, and it is possible to accidentally purchase one of these pseudoscientific tracts while searching for information about the Moon. If you go shopping for lunar books, *caveat emptor!*

The Moon's Motions and Environment

The Moon is a strange, fascinating place. The tenuous atmosphere of its surface is a near-perfect vacuum and so is perpetually quiet; no weather affects its terrain. The lack of an atmosphere means that the sky is perpetually black. Stars are visible from the surface during daytime but are difficult to see because the glare reflected from the surface dilates the pupils. At high noon, the surface can be over 100°C and, at midnight, as low as −150°C. The Moon is much smaller than Earth: its radius is only about one-fourth of Earth's and its mass is a little over 1 percent of Earth's. In surface area, the Moon is roughly the size of the continent of Africa, about 38 million sq km. Its day (the time it takes to rotate once on its spin axis) is about 29.5 Earth days or 709 hours, and daylight hours on the Moon (sunrise to sunset) last about two weeks. The Moon is famous for its low gravity, about one-sixth of Earth's. Thus, an astronaut who weighs 200 pounds on Earth weighs only 34 pounds on the Moon. However, mass remains the same, meaning that it still takes the same effort to get going and to stop (as several fall-on-your-face astronauts discovered during the moonwalks!).

The Moon orbits Earth, showing a different *phase,* or lighting conditions of its surface, at various times of the month (Fig. 1.4). The orbit of the Moon is elliptical, and it completes its circuit once every 27.3 days, a period known as the *sidereal month* (Fig. 1.5). Because Earth is moving around the Sun, the sidereal month is slightly shorter than the time between successive new moons, that is, the time for the Moon to rotate once on its axis

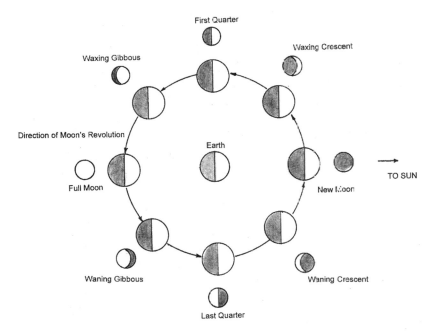

Figure 1.4. The Moon's circuit. The Moon changes phase because of its revolution around Earth; when it gets between Earth and the Sun, it is a new moon; when Earth is between the Sun and the Moon, we have a full moon.

(the lunar day). This period is known as the *synodic month* and averages 29.5 days long (Fig. 1.5). A consequence of this orbital phasing is that the Moon shows the same hemisphere (called the *near side*) to Earth at all times. Conversely, one hemisphere is forever turned away from us (the *far side*). You may hear the *dark side* of the Moon discussed, and there is indeed a dark side—it is the hemisphere turned away from the Sun, that is, the nighttime hemisphere. As such, the position of the dark side changes constantly, moving with the dividing line between sunrise and sunset (which is called the *terminator*). In some minds the far side has become confused with the dark side, but you should mark the distinction. The two probably became mixed up because of the cultural equation of "dark" with "unknown" (as in "darkest Africa," which is not really dark but was largely unknown to most of Europe in the early 19th century). Before the space age,

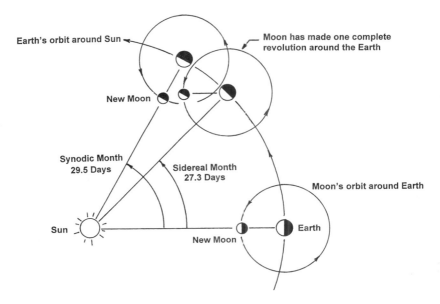

Figure 1.5. Diagram of the Moon's and Earth's orbits. Earth is moving around the Sun, resulting in a difference between the time it takes for the Moon to make one revolution around Earth (the *sidereal month,* or the time between successive star alignments, 27.3 days long) and the time it takes for the Moon to rotate once on its axis (the *synodic month,* or lunar "day," averaging 29.5 days or 709 hours).

the far side of the Moon was completely unknown territory, not revealed to human gaze until first photographed by the Soviet spacecraft *Luna 3* in 1959.

The elliptical orbit of the Moon results in a variable distance between Earth and the Moon. At perigee (when the Moon is closest to Earth), the Moon is a mere 356,410 km away; at apogee (the farthest position), it is 406,697 km away. This is different enough that the apparent size of the Moon in the sky varies; its average apparent size is the same as that of a dime held at arm's length. In works of art, a huge lunar disk looming above the horizon is often depicted, but such an appearance is an illusion. A moon near the horizon can be compared in size with distant objects on the horizon, such as trees, making it seem large, whereas a moon near zenith (overhead) cannot be compared easily with earthly objects and, hence, seems smaller.

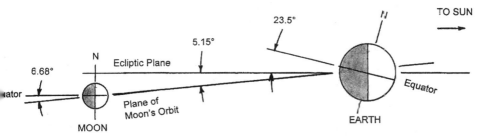

Figure 1.6. The plane of the Moon's orbit. Whereas Earth's spin axis
is tilted almost 24°, the Moon's axis is nearly vertical (1.5°); this
means that there are no appreciable "seasons" on the Moon.
However, the Moon's orbital plane is inclined to Earth's equator,
allowing us to look into and beyond the polar regions to the far side.

The plane of the Moon's orbit lies neither in the equatorial
plane of Earth nor in the ecliptic plane, in which nearly all the
planets orbit the Sun (Fig. 1.6). This relation poses some constraints on models of lunar origin. The spin axis of the Moon is
nearly perpendicular to the ecliptic plane, with an inclination of
about 1.5° from the vertical. This simple fact has some truly
significant consequences. Because its spin axis is vertical, the
Moon experiences no "seasons," as does Earth, whose inclination
is about 24°. So as the Moon rotates on its axis, an observer at
the pole would see the Sun hovering close to the horizon. A large
peak near the pole might be in permanent sunlight while a crater floor could exist in permanent shadow. In fact we now know
that such areas exist, particularly near the south pole (Fig. 8.6).
The existence of such regions has important implications for a
return to the Moon.

As the Moon circles Earth, the two occasionally block the
Sun for each other, causing an *eclipse*. A *solar eclipse* occurs
when the Moon gets between the Sun and Earth (Fig. 1.7) and
can occur only at a new moon (the daylit, or illuminated, side
of the Moon is facing the Sun). Because of the variable distance
between Earth and the Moon, the Moon's inclined orbital
plane, and the smaller size of the Moon, solar eclipses are quite
rare (years may pass between total solar eclipses), so an occurrence is always subject to much hoopla. A *lunar eclipse*, in contrast, occurs when Earth gets between the Moon and the Sun

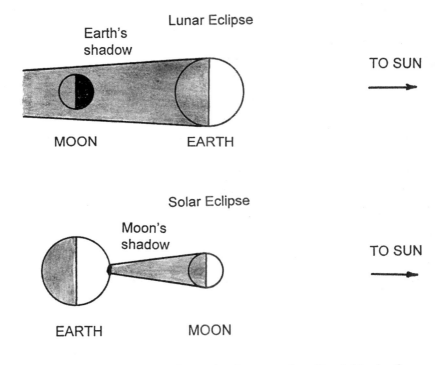

Figure 1.7. Eclipses. Lunar eclipses (top) occur when Earth blocks the light from the Sun; they thus can happen only during a full moon. In contrast, a solar eclipse (bottom) occurs when the Moon gets between the Sun and Earth; this can happen only during a new moon.

(Fig. 1.7). These events happen much more frequently because Earth's shadow has a much larger cross-sectional area than does the Moon's shadow. Lunar eclipses can occur only during a full moon (or new Earth). As the shadow of Earth slowly covers a full moon, it takes on a dull red glow, caused by the bending of some sunlight illuminating the Moon through the thick atmosphere of Earth.

The Moon is gradually receding from Earth. Early in planetary history, Earth was spinning much faster and the Moon orbited much closer than now. Over time, energy has been transferred from Earth to the Moon, causing the spin rate of Earth to decline and the Moon to speed up in its orbit, thus moving farther away (the current rate is about 4 cm/year). Such recession will continue; someday the Moon will be too far away to create a total solar eclipse. Fortunately for lovers of cosmic spectacles, this will not happen for at least another few million years.

As the Moon orbits Earth, we can peek around its edges because of a phenomenon known as *libration* (Fig. 1.2). Libration in latitude is caused by the 7° inclination of the plane of the Moon's orbit to Earth's equator. This inclination allows us to "look over the edge" of the Moon as it moves slightly above or slightly below the equatorial plane. Libration in longitude is caused by the Moon's elliptical orbit, which permits Earth viewers to look around the Moon's leading or trailing edge. A small libration is also caused by parallax, which is the effect that allows you to see more by moving side to side, in this case by the diameter of Earth. All told, these libration effects permit us to see slightly more than a single hemisphere, and over the course of time we can see about 59 percent of the lunar surface.

Gravity makes the solar system go around, and the gravitational influence of Earth and the Moon on each other is considerable. Because of the gravitational tug of the Moon and the Sun, Earth experiences *tides*, which are bulges in the radius of Earth induced by gravitational attraction. Tides are often thought to be associated with the oceans, but the solid Earth also undergoes an up-and-down motion caused by the tides. Tides can be exceptionally large, as when Earth, the Moon, and the Sun are in alignment (Fig. 1.7), a condition known as *syzygy* (a great word for a crossword puzzle!). Because Earth also attracts the Moon, the Moon too experiences a tidal bulge, one that mirrors the tidal effects on Earth. The raising and lowering of solid body tides on Earth and the Moon causes friction inside the two planets, a source of heat called *tidal dissipation*. Such an energy source for planetary heat may have been very important early in the history of the solar system, when the Moon and Earth were closer together, but is currently only a minor source of heat.

The Beginnings of Lunar Science

In this book I consider lunar science to be synonymous with the study of the Moon's origin and evolution as a planetary body. After Isaac Newton developed a comprehensive explanation of the Moon's orbit and motions with the publication of his *Principia* in 1687, the scientific study of the Moon began in earnest when people considered the origin of its surface features, specifically the craters, the dominant type of landform on the surface.

Although much effort was expended trying to determine the nature of the Moon's surface materials (mostly by studying the properties of the light reflected from the surface), the history of lunar science is largely the history of the debate on the origin of craters.

The word *crater*, of Greek origin (κρατερ), means cup or bowl. The Moon's surface is covered by craters of all sizes, as small as the limits of visibility and as large across as a small country. Proposed origins for craters fall into two categories: processes external to the Moon, such as the collision of asteroids and meteorites with the Moon, or internal processes, such as volcanism or collapse. One of the first scientific models was advocated by the 17th-century English scientist Robert Hooke, who conducted experiments with boiling alabaster to suggest that craters are frozen, partly burst "bubbles" created by the slow release of gas from a "boiling" surface. Hooke's model implies that the Moon either is or was at one time molten.

Over the subsequent 200 years, volcanism of one type or another seems to have been astronomers' favorite mechanism for crater formation. Astronomers were the principal type of scientist to study the Moon, mostly because they used telescopes, which was how the Moon was observed. They were impressed by the supposed resemblance between lunar craters and a type of crater found at the summit of many terrestrial volcanoes, *calderas*. However, in 1876 the British astronomer Richard Proctor, proposing that craters resulted from the collision of solid bodies with the Moon, outlined what came to be called the *impact model* of crater formation. Proctor's proposal was not widely embraced. Partly because no one had yet seen and described an impact crater on Earth, it was considered too exotic a model to be thoroughly tested. Why invoke a process no one had ever seen when dozens of clearly volcanic craters exist on Earth? Although he contended later in his life that he had not abandoned the idea, Proctor toned down his suggestion in subsequent editions of his book *The Moon* (the idea is completely missing from the second edition of 1878). Proctor's book also contained some of the first "realistic" imaginative scenes of what the Moon might look like to a person standing on its surface (Fig. 1.8), a view quite different from the imaginings of the fantasy writers.

Figure 1.8. Drawings of the lunar surface presented by Proctor in 1878. Proctor tried to portray the surface as accurately as science would permit, showing a dry, barren landscape with a black sky and with Earth shining brightly above.

In 1892 the American geologist Grove Karl Gilbert became interested in the Moon. In retrospect Gilbert's work seems very modern, even though his advocacy of the impact model still suffered from the lack of a suitable terrestrial example for comparison. Knowing that meteorite fragments had been found in the vicinity of Meteor Crater (then known as Coon Mountain) in northern Arizona, Gilbert studied the area as a possible example of how an impact could create a hole in the ground. After field study, Gilbert concluded that Meteor Crater had been formed not by impact but by a steam explosion! He arrived at this erroneous conclusion because the mechanics of high-velocity impact were not understood in his day. Gilbert believed that if impact had been the cause, a huge buried meteorite should exist beneath the floor of the crater, and his magnetic survey failed to reveal the presence of any significant iron meteorite body. We now understand that this absence results from the near-complete vaporization of the projectile on impact. The meteorites found scattered around the crater were largely split off from the main impacting mass during its passage through the atmosphere.

Undeterred, Gilbert still thought that most of the craters on the Moon were of impact origin. He based his argument on the great size of some of the craters, including the gigantic basin-sized craters that Gilbert first recognized, and on the fact that lunar craters do not resemble volcanic calderas in any fashion. Gilbert also made small, experimental craters by dropping balls of clay and shooting a pistol into clay and sand targets. What puzzled him most about the Moon's craters was their high degree of circularity. Gilbert computed that the average angle of impact would be around 45° for cosmic bodies hitting the Moon. He reasoned that this would result in a great many oval or elliptical craters; although some are seen, they are rare. Gilbert rationalized this rarity away by assuming that the craters of the Moon had been formed very early in history, when the Earth-Moon system had a ring of debris orbiting both bodies, like the rings of Saturn. Again, the crude state of knowledge of impact processes betrayed Gilbert; we now know that craters do not become elliptical in outline until the angle of impact becomes lower than about 5° above the horizontal.

Gilbert's last contribution to lunar science was significant indeed. He was the first to recognize that the circular Mare Imbrium was the site of a gigantic impact and that a pattern he called *sculpture* was the surface trace of material hurled out of the basin and spread across the entire near side (Fig. 1.9). Because he could detect craters that formed both before and after that event, Gilbert even suggested that a scheme to classify surface features by *relative age* could be built around the *ejecta* of the Imbrium *catastrophe*. This recognition that the Moon is a complex body, built up by innumerable events over a long period of time, was the key to unraveling the history of the Moon. Curiously, Gilbert did not pursue his suggestion, and it awaited the coming of the space age before we would map the geology of the Moon (see Chapter 3).

Gilbert's fascination with impact as a geological process was not unique. In the early 20th century Daniel Barringer, a mining engineer, embarked on a decades-long struggle to prove that Gilbert was wrong about the origin of Meteor Crater. He was convinced that the crater *had* been formed by impact and that a great fortune could be made by mining the nickel from the massive nickel-iron meteorite that he was sure was buried beneath the crater floor (thus he repeated Gilbert's principal mistake). Barringer ultimately went broke, investing all of his assets into a program of drilling the crater floor, looking for the buried meteorite. However, Barringer's campaign for the impact origin of Meteor Crater kept the idea alive when most scientists were ready to abandon the concept.

In 1921 the German geologist Alfred Wegener, best known as the father of the concept of continental drift, wrote a lengthy defense of the impact origin of the Moon's craters. Fourteen years later the American geologists John Boon and Claude Albritton wrote an intriguing article suggesting that several very large, circular structures on Earth are in fact eroded impact craters, features they called *astroblemes* (meaning "star wounds"). This suggestion opened up a whole new field of inquiry; remember that one of the principal problems with the impact idea was the lack of a clear example on Earth. Some of the best examples of very large craters on Earth are in Canada, where the stable heartland of the North American continent has preserved many old impact craters. With Barringer's champion-

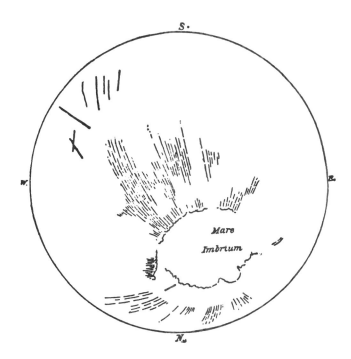

Figure 1.9. Top: a drawing of the near side of the Moon by Gilbert (1893), showing the radial pattern surrounding the Imbrium basin, a pattern he called "sculpture." Bottom: an Apollo photograph of the central highlands south of the Imbrium basin showing the sculpture, which is caused by material thrown out of the basin during the impact.

ing of Meteor Crater and with the collection of examples cited by Boon and Albritton, various craters of different ages, sizes, states of erosion, and shapes were now available for geological examination. The active search for old impact craters on Earth continues to this day. Workers have discovered over 100 impact craters on Earth and believe that many more remain to be found. The latest (1991) spectacular discovery was that of the buried, giant Chicxulub Crater in Mexico, almost 300 km in diameter and thought to be the impact responsible for the extinction of the dinosaurs (and many other life forms) 65 million years ago.

The impact cause was picked up by geologist Robert Dietz in a paper published in 1946. Dietz argued that almost all lunar craters are impact craters, and he set out to prove this concept by studying many impact craters on Earth. One of the most interesting impact features is the Sudbury structure in northern Ontario, long famous as a nickel-mining area. Dietz, who thought that the area might also be a giant impact basin, ultimately found the first evidence for impact at Sudbury in the form of shatter cones, small grooved structures in rock caused by the passage of a shock wave (something created only in nuclear explosions and cosmic impacts). As more craters continued to be discovered and studied on Earth, the case for an impact origin for lunar craters became ever more compelling. It remained only for someone to put all of the pieces together. In 1949 someone did.

The Pre-Apollo Moon

Ralph Baldwin is an astronomer and physicist by training and a businessman by profession. He is also a lifelong student of and enthusiast for the Moon. In his 1949 book *The Face of the Moon*, Baldwin got nearly everything right. Baldwin noted the similarity in form between craters on the Moon and bomb craters on Earth, created in abundance during World War II. On the basis of telescope observations and meteor entry data, he also plotted the number of meteoroid objects in space as a function of size (called the *size-frequency distribution*) and determined that the shape of that curve perfectly matched the size-frequency distribution of craters on the Moon, a strong argument that the craters were formed by impact. Baldwin did not advocate that every lunar feature is of impact origin: he thought that the dark, smooth maria are flows of basalt lava, similar to flood lava plateaus on Earth. Finally, his work documented that the circular maria are actually very large craters, craters that were originally created by impact and filled later with lava (when writing his book, Baldwin was unaware of Gilbert's 1893 paper, which was published in a relatively obscure journal, or of Dietz's more recent [1946] effort). Recognition of the impact origin of these features (we now call them *basins*) was a major step toward understanding the importance of these features for the geological evolution of the Moon.

A contrasting view of lunar history was offered by Harold Urey, a chemist who had won the Nobel prize in 1934 for his discovery of heavy hydrogen. Urey became interested in the Moon after reading Baldwin's book. He claimed that because the Moon now appeared to be a cold, rigid body, it had always been thus and that, therefore, almost all craters are of impact origin (true), the maria are debris blankets from the basin impacts (false), and the Moon is a cold, primitive body that never underwent any significant melting (very false). Because of Urey's immense scientific stature, his views on the Moon were taken seriously, and his vocal advocacy for the Moon was a significant contributing factor in making the Moon an early goal of the infant space program.

The man who really got the geological ball rolling toward the Moon is still actively pursuing lunar and planetary science. Carefully following the army's experiments with captured German V-2 rockets in the late 1940s, Eugene Shoemaker realized that people soon would travel to the Moon and that, since it was obvious to him that the Moon is a body shaped by geological processes, a geologist would have to be sent to conduct the exploration of the Moon. He fully intended to be that geologist. Shoemaker founded the Branch of Astrogeology of the U.S. Geological Survey (USGS) in 1961. His careful analysis showed that the Moon's surface could be studied from a geological perspective by recognizing a sequence of relative ages of rock units near the crater Copernicus on the near side (this area is discussed in the next chapter). Shoemaker also carefully studied Meteor Crater in

Arizona and documented the impact origin for this feature, successfully completing the initial quest of G. K. Gilbert some 60 years earlier. You may be familiar with this productive and energetic scientist by the recent acclaim that he and his wife, Carolyn Shoemaker, received from their discovery of the Shoemaker-Levy comet, which hit the planet Jupiter in 1994, right before the eyes of an incredulous world. (We had never before actually observed a giant impact.) Today, in an effort to better understand the rates of impact on the planets, Gene and Carolyn continue their search for impacting debris.

Under Shoemaker's guidance, the USGS began to map the geology of the Moon using telescopes and pictures. Along with the work of Baldwin, this mapping, conducted during the 1960s, effectively gave us our pre-Apollo understanding of the Moon. We became more familiar with the impact origin of craters and basins, ascertained the relative ages of surface features (including the ancient highlands, the younger maria, and the very youngest rayed craters), and made considerable progress toward comprehending the important geological processes, such as volcanism, impact, and tectonism. In 1971 the USGS scientists Don Wilhelms and Jack McCauley published a geological map of the entire near side that shows the relations between geological units. This map was a synthesis of the work done to prepare for Apollo and can be profitably studied by lunar students even today, although some interpretations have been superseded by the findings of Apollo.

The stage had been set. For hundreds of years, people had pondered and wondered about the Moon. The emergence of the technology of rocket propulsion and spaceflight opened up new vistas for exploration. Because the Moon is the closest planetary body to us, it was a logical target for our first, faltering steps into the solar system. All that was needed now was an appropriate trigger to provide a suitable financial and political climate whereby millions of research and development dollars might become available. In May 1961 the trigger was pulled.

 Chapter 2

The World of the Moon

Before we look in detail at the exploration of the Moon and its history as inferred from that activity, we should acquaint ourselves with its surface features. The radius of the Moon is 1,738 km, whereas the radius of Earth is 6,371 km; thus the Moon is relatively large in relation to its parent planet. The satellites of the outer planets are very tiny compared with their hosts, and the Moon could be considered to be a small planet. As we shall see, the Moon has indeed undergone a distinct and complex geological history, and much of our understanding of the terrestrial planets, especially Mars and Mercury, derives from or is constrained by our knowledge of the Moon. In this sense the Moon is indeed a Rosetta stone that allows us to read the otherwise indecipherable text of planetary evolution written in the crusts of the terrestrial planets.

A Scarred and Jagged Landscape: The Moon's Face

Through a telescope, the Moon presents a striking appearance (Fig. 2.1). The Moon may be divided into two major terrains: the *maria* (dark lowlands) and the *terrae* (bright highlands). The contrast in the reflectivity of these two terrains suggested to many early observers that the two terrains might have different compositions, and this supposition was confirmed by the Surveyor and Apollo missions. One of the most obvious differences between the two terrains is the smoothness of the maria in contrast to the rough highlands. This roughness is mostly caused by the abundance of craters; the highlands are completely covered by large craters (greater than 40–50 km in diameters), while the

Figure 2.1. View of the Moon from a telescope on Earth. Note the smooth, dark lowlands *(maria)* and the bright, rough highlands *(terrae)*. The prominent rayed crater at bottom is Tycho (see Fig. 2.4).

craters of the maria tend to be much smaller. After an extended debate lasting over 200 years, we now know that the vast majority of the craters on the Moon are formed by the impact of solid bodies with the lunar surface (see Chapter 1).

Most of the near side was thoroughly mapped and studied from telescopic pictures in the years before the space age. Earth-based telescopes can resolve objects as small as a few hundred meters on the lunar surface. Craters were observed, down to the smallest sizes. This observation, and the way the Moon diffusely reflects sunlight, led to our understanding that the Moon is covered by a ground-up surface layer, or *regolith*. Telescopic images permitted us to catalog a bewildering array of landforms. Craters were studied for clues to their origin. The large circular maria were mapped. Wispy marks on the surface (rays) emanating from certain craters were seen. Strange, sinuous features were observed in the maria. Although various landforms were cataloged, the attention of the majority of workers was (and still is) fixed on craters and their origins.

Early in the history of lunar studies, it was noted that the shape of craters changes with increasing size. Small craters with diameters of less than 10–15 km have relatively simple shapes (Fig. 2.2). They have rim crests that are elevated above the surrounding terrain, smooth, bowl-shaped interiors, and depths that are about one-fifth to one-sixth their diameters. A rough-surfaced deposit surrounds the rim crest of these craters and extends out to a range from about a crater radius to a crater diameter. A multitude of tiny, irregular craters are found beyond the outer edge of the textured deposit, surrounding the host crater like a swarm of bees. The textured deposit is termed *ejecta* and consists of material that was thrown out of the crater during its formation. This crater ejecta is continuous near the rim crest of the host crater but becomes discontinuous farther out. The tiny irregular craters are called *secondary craters*. They form when clots of debris heaved out from the main crater hit the surface. As you might expect, secondary craters greatly outnumber primary craters on the Moon; the collision of an asteroid will create only one primary crater but a multitude of secondaries.

The complexity of shape increases for larger craters. Larger, bowl-shaped craters in the diameter range of 20 to 30 km have flat floors and scalloped walls. Craters in this size range (Fig.

Figure 2.2. The bowl-shaped crater Linné (2 km diameter) in Mare Serenitatis. The smallest craters on the Moon show this simple shape. The bright zone around the crater is its ejecta blanket, made up of crushed rock and debris thrown out of the crater.

2.3) straddle the divide between what are called *simple craters* (bowl-shaped) and *complex craters* (craters with wall terraces and central uplifts). Around 30–40 km diameter, we begin to see central peaks and wall terraces (Fig. 2.4). Terraces are easily understood as the logical continuation of wall scalloping, but the appearance of central peaks is somewhat unexpected. It might be that the pooled wall debris found in the floors of bigger craters is merely piling up, creating a peak. However, studies of the central peaks of complex craters on Earth show that these peaks are not surface features but reflect deep-seated structural deformation. The central peaks of complex craters may come from depths as great as 10 to 20 km in the Moon. As such, they are invaluable because they allow us to look at the composition of the crust at depth.

One of the most beautiful complex craters on the Moon is

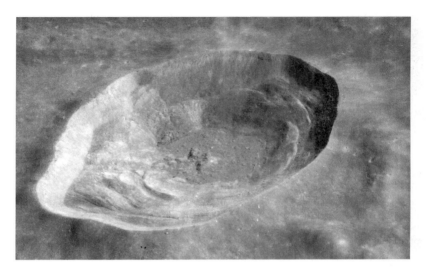

Figure 2.3. A 20-km-diameter crater in Mare Nubium. In this size range, wall failure results in large slumps, or wall scallops, that make up pools of debris on the crater floor. Small central uplift is evident.

Tycho (Fig. 2.4; Plate 2). At 87 km in diameter, it is roughly the size of the metropolitan area of Los Angeles. This feature has one of the most spectacular ray systems on the Moon, extending in some cases over 3,000 km from its rim. Tycho displays all of the features of a complex crater in spectacular clarity because it is also one of the very youngest large craters on the Moon. *Apollo 17* sampled fragments of ejecta from Tycho's secondary craters, so we know the age of this feature: 108 million years (seemingly very old but young by lunar standards). The rim, walls, and floor of Tycho show fantastic, contorted shapes (Fig. 2.4); the surface texture closely resembles surface textures seen on very fresh lava flows on Earth. These materials are made up of a shock-melted sheet of liquid rock created when a large asteroid strikes the Moon. This *melt sheet*, which spreads out over the floor as the crater is formed, cools in place after the creation of the central peak and after the collapse of the rim to form the wall terraces. We know from study of impact craters on Earth that this melt sheet is created by the whole-scale melting of the target rocks and that, as such, it represents an average composition of the crust beneath the crater.

Figure 2.4. The crater Tycho (85 km diameter) in the southern, near-side highlands. Craters of this size display a central peak, flat floors, and wall terraces. Rough, lava-like floor materials are made of shock-melted rock from the impact.

Moving out from Tycho's rim crest, we see a rough, mountainous unit made up of crushed and broken rocks ripped out of the crater cavity by shock pressure. This deposit, called the crater *ejecta blanket*, covers the terrain out to a distance of up to 100 km. Beyond this distance, we can still detect patches of debris from Tycho, but the dominant surface features in this area are the secondary craters. Secondaries come in a range of shapes and sizes and are often clustered or aligned into rows (Fig. 2.5). The secondary crater field is made when large blocks, clumps of loosely bonded ejecta, and fine sprays of ground-up rock strike the Moon, possibly after journeys that may have taken them on flights up and down, through thousands of kilometers in space.

Figure 2.5. Small secondary impact craters and rays from the crater Copernicus (93 km diameter), just on the horizon. The secondaries form crude, V-shaped clusters that point back to the primary crater. Secondaries form by the impact of large clots of debris ejected from a primary crater.

The rays of Tycho (Fig. 2.1) extend almost 3,000 km away from the crater, and several extend well beyond the near to the far side. Rays around large craters consist of two parts: fine, ground-up rock powder from the crater cavity and bright local material thrown out from a secondary crater. Thus rays are made of material flung hundreds of kilometers across the Moon as well as material from the immediate locality.

Up to a much larger size, about 200 to 300 km in diameter,

Figure 2.6. The peak-ring crater Compton (185 km diameter), near the north pole. Craters in this size range show most of the features of complex craters, in addition to the ringlike elements of impact basins.

craters seem to maintain the shapes typified by Tycho. In larger craters, central peaks become more subdued and are made up of groups of peaks separate from each other. The most striking change is the appearance of a ring of peaks, in addition to the central peaks, as seen in the crater Compton (Fig. 2.6). The appearance of a ring bespeaks the next transition in crater shape, that of the change from crater to basin. The term *basin* has several meanings, but as used in this book, a basin is merely an impact crater larger than 300 km in diameter. The double-ring basin Schrödinger (Fig. 2.7) is a conspicuous, fresh basin near the south pole. At diameters greater than the 320-km diameter of this basin, no central peaks are evident.

The last major change in crater shape is the appearance of concentric, multiple rings, as in the spectacular, multiringed Orientale basin, almost 1,000 km across (Fig. 2.8). The units of the Orientale basin offer a guide to interpreting the bewildering complexity of the terrae. Multiring basins are the largest craters on the Moon; some of these features are over 2,000 km in diame-

Figure 2.7. The two-ring basin Schrödinger (320 km diameter).
Basins are craters whose central features have ring, rather than peak,
form. Older basins with no interior rings are presumed to have once
had them; if the rings are not evident, it is supposed that they are
buried under basin fill materials.

ter (as large as the entire western United States). The very largest basins may even have stripped off the entire crust at several
sites. Basins serve as regional depressions that accumulate mare
lavas and control the trends and distributions of faults and folds
on the surface. Thus, in many ways, basins make up the basic
geological framework of the Moon.

Basins and the Highland Crust

Orientale provides an example of a fresh, multiring basin (Fig.
2.8). Because it is the youngest basin, it is the only nearly unmodified example of its type on the Moon. Like complex craters,
basins are surrounded by deposits that were laid down at the
time of impact. The highland material in the interior of the
basin looks like the shock-melted sheet found at Tycho. Orientale is surrounded by a huge sheet of debris, thrown out of its

Figure 2.8. The spectacular Orientale basin (930 km diameter). This feature is one of the best-preserved lunar basins and has been used as an archetype for other, more degraded basins. The Orientale basin shows at least four (possibly five) rings, including three interior and one exterior to the basin rim crest.

excavated cavity over nearly a quarter of a hemisphere. The surface appearance of this blanket of debris changes systematically as a function of distance from the basin rim. These changes are probably related to the energy with which the material was thrown out, with the highest-energy debris traveling the farthest from the basin rim.

The interior of the Orientale basin is only partly flooded by later volcanic deposits (Fig. 2.8), so we can see the floor of a nearly unmodified impact basin. The floor displays smooth plains, rough, knobby deposits, and intensely fractured, undulating plains (Fig. 2.9). Collectively, these deposits resemble the rough, lava-like melt sheet on the floor of Tycho (Fig. 2.4), although on a much larger scale. We believe that these features are part of the huge sheet of shock-melted rock created during the impact. The smooth plains near the basin center represent the thickest, purest layer of the melt sheet, covering all of the hills and mountains of the inner basin floor. As we move out toward the rim, the melt deposit becomes thinner and mixed with unmelted debris. The subsurface hills and valleys become visible beneath a cracked and fissure-covered melt layer (Fig. 2.9). The rough, knobby material found just inside the basin rim (Fig. 2.9) has no comparable feature at Tycho and is probably some type of ejecta from the impact cavity. The knobby material may or may not contain impact-melt bodies ejected from the crater cavity.

Outside the rim, ejecta from the basin has a rough, hilly appearance. At places, a wormy texture is evident (Fig. 2.9), probably caused by the flow of material on the ground. Such surface flow occurs after the ejecta is lobbed out of the basin and travels through space on a ballistic path, as does the ejecta from smaller craters, such as Tycho (Fig. 2.4). The wormy textured units tend to occur within the floors of preexisting craters, suggesting that this material has piled up upon itself at the end of flow along the surface. The textured ejecta blankets of basins extend out proportionally from their rims, as do the ejecta blankets of craters; continuous basin ejecta from Orientale may be found up to 1,000 km from its rim crest.

Farther out, basin ejecta becomes more discontinuous, fragmentary, and isolated. Smooth, light-toned plains fill many of the older craters of the highlands at the farthest ends of the Orientale ejecta blanket (Fig. 2.8). These plains (called *highland*

Figure 2.9. High-resolution views of the interior of the Orientale basin (Fig. 2.8). The floor of the basin shows smooth and cracked materials that probably represent the impact melt sheet of the basin. The braided and knobby materials make up the ejecta blanket of the basin. All of these features are scaled-up versions of similar features found around smaller complex craters.

plains or *Cayley plains*, after a crater where they are well developed) form level surfaces and fill depressions. At one time these plains were thought to be ancient lava flows in the highlands. The results of the *Apollo 16* mission (see Chapter 3) showed that the Cayley plains are instead made up of impact breccias and are probably formed of far-flung ejecta from one or more of the very largest basins. Vast fields of large (5–20 km in size), irregular craters and crater clusters are associated with these smooth, light plains (Fig. 2.8). Because we see such irregular craters around the edges of the ejecta blankets of large complex craters, we believe that these irregular features in the highlands are secondary craters from basins, their large size reflecting the great size of a basin-forming impact. Basin secondaries tend to concentrate near the edges of the ejecta blanket. However, they can be found at great distances from the basin rim and may even occur on the side of the Moon directly opposite a basin (the antipode).

A variety of mysterious grooved, ridged, and hilly terrains is found in many places of the highlands. It was thought before the Apollo missions that these features, in conjunction with the strange irregular craters described above, were caused by volcanism in the terrae. We now understand that almost all of these features are related somehow to the process of basin formation. For example, a strange grooved texture of crater walls near the center of the far side, near the crater Van de Graaff (Fig. 2.10), is directly antipodal (opposite) to the center of the large Imbrium basin on the near side. This unit is not the result of volcanism in the highlands but is caused either by huge seismic waves traveling through the Moon from the basin impact, causing massive landslides as they converged near the antipode, or by the accumulation of very far-flung clots of ejecta thrown out by the impact creating the Imbrium basin. In either case the texture is caused by impact, not volcanic, processes.

Because most maria are found on the near side and these lavas fill in depressions in the crust, it was once thought that basins occur preferentially on the near side. However, basins cover the Moon equally on both the near and the far sides (Fig. 2.11). Most basins on the Moon, particularly those found on the far side, have little or no fill of mare basalt. The reasons for this paucity of mare filling in basins on the far side are unclear but may be related to regional variations in the thickness of the crust (see Chapter 6).

Figure 2.10. The area near the crater Van de Graaff, antipodal to the Imbrium basin. The unusual grooved texture seen on the rims of craters may be caused by the converged focusing of intense seismic waves, generated by the giant impact at the antipode of the basin.

The complete saturation of the Moon by very large basins means that the ejecta blankets of these features dominate the upper few kilometers of material in the highlands. In conjunction with the mixing of this same zone of the crust by the large population of craters, a large, thick layer of shattered and crushed rock is found in the outer few kilometers of the lunar crust. This giant zone of breccia is called the *megaregolith*, a term derived from the name regolith, given to the much smaller pulverized zone found on both mare and highlands (see Chapter 4). The presence of this megaregolith means that rocks returned from the highlands are all processed by impacts and that the original crust must be carefully reconstructed by studying tiny fragments that have managed to survive this bombardment.

One might think that this shattering and dislocation would

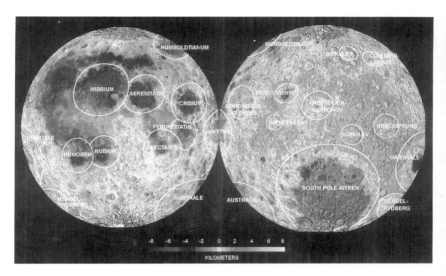

Figure 2.11. Map showing the location of recognized impact basins on the Moon. Basins are randomly distributed across the Moon; they are not preferentially concentrated on the near side, where most of the maria are found.

erase any primeval variations in the composition of the crust. Apparently this did not happen, because we see evidence for many different provinces of rock types on the Moon from sample and remote-sensing data. It appears that the effect of most of this bombardment is to crush, remelt, and break up the original crust but not to homogenize its composition. We know from the study of impact craters on Earth that only the shock melt created during a large impact homogenizes the target compositions. Thus a first-order conclusion from the lunar highlands is that the crust has not been completely remelted since its original creation but only remelted locally and in a few places, namely, the sites of the crater and basin melt sheets. In particular, basin-forming impacts, being very large events, instantly created huge volumes of shock-melted rock, and many of the melted breccias returned by the Apollo missions are thought to have been formed in basin impacts (see Chapter 6).

One other aspect of basin geology is important. As noted previously, basins are remarkable in their display of multiple, concentric rings. These rings are displayed most spectacularly at Orientale (Fig. 2.8), where at least four distinct basin rings are visible. The fourth ring, the Cordillera Mountains, makes up the basin rim and probably corresponds to the rim crest of complex craters, such as the rim of Tycho (Fig. 2.4). Moreover, these rings seem to be spaced at regular intervals. Such spacing occurs not only on the Moon but on the other terrestrial planets (including Earth) as well. Basin rings crisscross, intersecting and interfering with each other (Fig. 2.11). This complex network of rings also contributes to the patchwork geology of the highland crust.

The origin of basin rings is still something of a mystery. It looks like a very deep seated structure is associated with the basin rings because they are often the sites of eruption of volcanic lava from great depth. This association might indicate that the rings are giant fault scarps and that the fracture along which the scarp formed served as a pipeline for the movement of lava to the surface. Another possibility is that basin-forming impacts fluidize the surface and that rings are "frozen waves" in the crust, much like the ripples formed when a stone is thrown into the still water of a pond. As such, basin rings would be primarily features of the surface, not the deep crust, and would not necessarily be associated with deep fractures. It is also possible that rings have some type of composite origin, some rings forming by one mechanism and other rings by another mechanism.

Lava Flows and the Maria

The dark, lowland plains of the maria are very striking in appearance (Fig. 2.1), yet the maria seem to occupy a larger fraction of the surface than they actually do. In fact mare deposits make up only about 16 percent of the surface area of the Moon. At the scale of telescopic observation, the maria appear smooth and relatively uncratered, at least compared with the highlands. The landforms of the maria likewise tend to be small in relation to those of the highlands. The small size of mare features relates to the scale of the processes that formed them, that is, local volcanic eruptions as opposed to the regional- and global-scale effects of the basin-forming impacts.

Some of the most obvious features in the maria are the winding, blister-like ridges that snake across the surface (Fig. 2.12).

Figure 2.12. A wrinkle ridge in the maria. Wrinkle ridges are tectonic features that are formed by the surficial compression of a regional unit. You can think of them as "buckled" surface features.

These features are called *wrinkle ridges* and are found in virtually all regional mare deposits. Wrinkle ridges often follow circular trends inside a mare basin, align with small peaks that stick up through the maria, and outline the interior ring structure in basins deeply filled by mare deposits. Circular ridge systems, such as the spectacular ridge complex associated with the crater Letronne (Fig. 2.13), outline buried features, such as the rims of pre-mare craters. Wrinkle ridges are broad folds in the rocks, some possibly faulted, and are caused by the regional compression of the mare deposits (see Chapter 5).

Within Mare Imbrium, we can see a series of lobate scarps, winding their way across the surface (see Chapter 5). Because similar features are seen in volcanic fields on Earth, we think that these scarps are lava flow fronts. The scarps are one piece of

Figure 2.13. The partly flooded impact crater Letronne, in Oceanus Procellarum. Note that the "missing" part of the crater rim is outlined by wrinkle ridges; also, a median ridge marks the location of the (presumably) buried central peak.

a variety of evidence indicating that the maria are made up of lava flows of basalt, a dark, common rock found in abundance on Earth. Flow fronts are rare on the Moon. This rarity could be caused by the great age of the mare lava flows, with impact erosion having destroyed the topographic evidence for flow fronts except in a few places. Another possibility is that large flow fronts never developed, for most of the eruptions and the maria consist of very thin, multiple flow fronts. We find evidence on the Moon for both possible causes.

Snakelike, sinuous depressions, called *rilles*, meander their

Figure 2.14. The crater Krieger (19 km diameter), showing a volcanic infilling in a region with numerous sinuous rilles (lava channels). Sinuous rilles are found in many areas of the maria.

way across many areas of the maria (Fig. 2.14). Although it was once thought that these rilles could be ancient riverbeds, we know from the extreme dryness of the Apollo samples (see Chapter 3) and from details of their shape that sinuous rilles are probably channels formed by the flow of running lava, not water. Some rilles begin in the highlands, so erosion by the flowing lava is a possibility. However, we know from the study of lava channels on Earth that they typically form within a single lava flow and are mainly constructional, not erosional, features. On Earth, lava channels are sometimes covered by roofs of lava, forming a lava tube. After the eruption the tube may drain,

Figure 2.15. Small cones and domes near the crater Hortensius. These features are small, central-vent volcanoes that indicate relatively low rates of effusion of lava from a small central vent, allowing constructs to be built up. The shield shape is indicative of low-viscosity (very fluid) lava, such as basalt.

leaving behind a cave in the lava flow. This process raises the intriguing possibility that caves may exist, or may once have existed, on the Moon. If one could be found, a cave on the Moon would be an attractive location for human habitation, offering protection from the extreme radiation and thermal environment of the lunar surface (see Chapter 9).

Scattered throughout the maria is a variety of other features that probably formed by volcanism. Small hills with pits on top are probably little volcanoes (Fig. 2.15); such lava shields are often found in volcanic fields on Earth. Isolated domes that may represent small-scale, low-volume eruptions occur throughout the maria. Both domes and cones tend to cluster together at many localities, as do volcanic cones on Earth. One of the largest concentrations of volcanic cones on the Moon is the Marius Hills in Oceanus Procellarum. Over 50 separate central-vent volcanoes and numerous wrinkle ridges and sinuous rilles occur within this complex.

In addition to the dark plains of the maria, we find large areas covered by a very dark material that seems to blanket both maria and terrae (Fig. 2.16). This material, called a *dark mantle deposit*, is similar in appearance and distribution to the dark volcanic ash found near the summits of some volcanoes on Earth. Results from the Apollo missions confirmed that these deposits are indeed volcanic ash (see Chapters 3, 5). Other dark mantle deposits are much more localized, found in association with small craters on the floors of some large craters with fractured floors. These features may be cinder cones. The last type of dark mantle deposit is also associated with craters but with small impact craters on some plains of the highlands. These impact dark-halo craters excavate dark material from beneath a lighter surface layer. They are extremely significant for the volcanic history of the Moon, as we shall see in Chapter 5.

Certain craters in the maria are irregular in shape and do not resemble typical craters on the Moon. At least some of these craters are probably of volcanic, not impact, origins. They may be eruptive centers, the source craters for the effusion of lava, or they may be collapse pits, caused by the sudden withdrawal of lava from beneath a thin crust and the resulting collapse of the surface. It was once thought that some strange-appearing large craters were also of volcanic origin, such as the crater Kopff (Fig. 2.9). We now believe instead that most of these features are impact craters that formed under unusual conditions, such as an impact into a partly molten target. However, this idea does not explain all the occurrences of such craters. The full story of these unusual features has yet to be written.

Odds and Ends

Across the Moon, both in highlands and in maria, we find strange landforms that do not conform to our notions or understanding of lunar processes. Oddly shaped pits, depressions, and irregular dimples do not resemble the known landforms created by either impact or volcanism, the two dominant surface processes. Sometimes these strange landforms are associated with

Figure 2.16. The dark mantle deposits surrounding Rima Bode II, the cracklike feature seen near the center. These dark deposits are made up of tiny glass beads of basaltic composition; such material was sampled on the *Apollo 17* mission. It is the lunar equivalent of volcanic ash, caused by fire-fountain eruptions on the Moon.

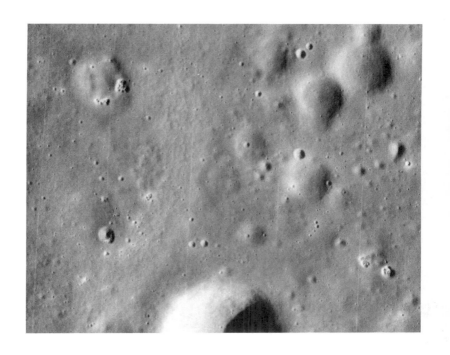

bright or dark patches (Fig. 2.17). Such areas may be places where pockets of gas have escaped out of the deep interior. Unusual craters (Fig. 2.17) could indicate an unrecognized type of secondary impact crater, or they could be manifestations of an atypical type of projectile, such as an extremely low density "fluffball" of ice and dust, as has been postulated for certain cometary debris. In some cases unusual shapes indicate an interplay of more than one process, such as an impact feature modified by volcanic eruption or structural deformation.

Various miscellaneous scarps and fissures occur all over the Moon. Of particular interest are the abundant small (less than 1 km across) scarps that appear to be randomly distributed throughout the highlands (Fig. 2.17). Because these scarps appear to overlie very fresh (young) impact craters, they have been thought to be ridges formed during a recent phase of crustal deformation. If such an interpretation is correct, the Moon is currently experiencing a global contraction of considerable magnitude. This hypothesis is controversial, and such a model awaits testing by a systematic survey of the entire Moon with high-resolution photographs, images that could be obtained on a future orbiting mission.

One unusual deposit on the Moon deserves special mention. Near the crater Reiner in Oceanus Procellarum is a bright, swirl-like deposit (Fig. 2.18), Reiner Gamma (Reiner ψ). Reiner Gamma appears to be a very young feature because very few craters can be shown to overlie it. These swirls are also coincident with a very strong (for the Moon) surface magnetic field, as detected by low-orbiting spacecraft. The Reiner Gamma swirls are not unique; similar deposits are found north of Mare Marginis on the eastern limb (Fig. 2.18) and near the crater Van de Graaff on the far side (Fig. 2.10). These later occurrences are near the antipodes of the Orientale and Imbrium impact basins, respectively. Like the

---

Figure 2.17 (opposite). Top: small, bright-rimmed pits that suggest relatively recent activity. These pits may represent sites where gas has escaped from the interior. Bottom: strange, bulbous domes. These features may represent either clumps of ejecta of unusual properties (e.g., melt) from a distant crater or extrusion of small amounts of lava from the interior. There are many unusual features on the Moon for which we have no good explanation.

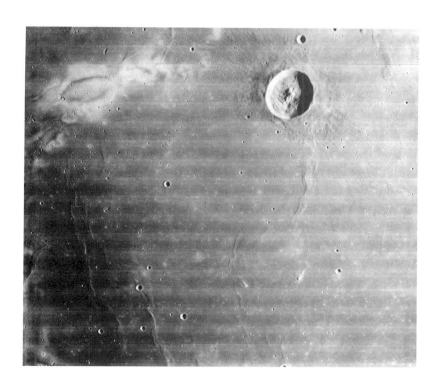

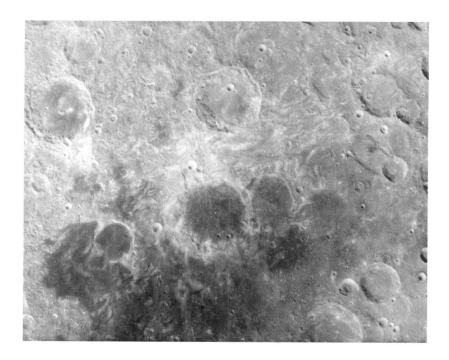

Reiner Gamma swirls, they also show strong surface magnetic fields.

We do not know how these swirls formed. One suggestion is that they are areas where volcanic gases have vented from the Moon, discoloring the local surface materials. Because the surface tends to darken with time as a result of bombardment by solar wind gases, another idea is that the Reiner Gamma swirls are normal surface materials that have been shielded from the solar wind darkening by their intense surface magnetic fields. Yet a third idea is that these swirls are the surface remnants of the impact of a comet. The hot plasmas and gases associated with the cometary nucleus scoured the surface materials (creating the light color) while the intense magnetic field of the nucleus induced the local magnetic anomaly. There are problems with each of these explanations, and the origin of the Reiner Gamma swirls must be added to the list of unsolved lunar mysteries. Their presence is mute testimony that the Moon retains its ability to perplex us.

Geological Mapping and Lunar History

Many people are surprised to learn that the basic outlines of lunar history were understood before we went to the Moon. A program to systematically map its geology had delineated most of the Moon's principal events after crustal formation. Geological mapping is done from photographs. It relies on an extremely simple foundation: Younger rocks and geological units overlie or intrude into older rocks and units. This principle is called the geological *law of superposition* and applies on the Moon, Earth, and in fact, all of the rocky objects of our solar system. How can such a conceptually simple statement decipher the complex history of a planet?

Figure 2.18 (opposite). Top: the Reiner Gamma swirl, a single, elongate, diffuse bright marking. Bottom: the Mare Marginis swirls, which are multiple and have diverse shapes. Both sites are associated with very strong magnetic fields. Their origin is unknown, although it has been proposed that they represent cometary impact sites. Bright swirls are found in several locations on the Moon.

The way we mapped the geology of the Moon can best be understood by looking at the area where the basic framework was first understood (Fig. 2.19). The beautiful crater Copernicus is a fresh, rayed crater, whereas its neighbor Eratosthenes shows no rays. Examination of detailed pictures shows that small secondary impact craters and rays from Copernicus overlie the rim of Eratosthenes, demonstrating that Copernicus is younger than Eratosthenes. Likewise, both craters are younger than the mare plains, on which both craters have formed. However, the crater Archimedes (Fig. 2.19) is filled by mare lava; therefore, it was created before the floods of surface lavas that created Mare Imbrium. Note that Archimedes occurs within the rim of the Imbrium basin, represented by the Apennine Mountains (Fig. 2.19). Clearly, both the mare plains and the crater Archimedes were laid down after the large impact that created the Imbrium basin. Thus, from these simple observations, we derive the following sequence of creation, from youngest to oldest: Copernicus, Eratosthenes, mare plains, Archimedes, Imbrium basin.

This region gave some of the names of its features to the scheme we now use to map lunar geology. Rayed, fresh craters that appear to be the youngest features in a given area are assigned to the Copernican System. Craters that have no rays, that are slightly eroded, but that mostly postdate the mare are part of the Eratosthenian System. The maria and older craters, both flooded and unflooded, are assigned to the Imbrian System, as are the Imbrium and Orientale basins. All deposits that formed before the Imbrium basin originally were classified as pre-Imbrian, but we have now subdivided that informal system into the Nectarian System (materials deposited between the impacts that created the Nectaris basin and the younger Imbrium basin) and the pre-Nectarian (all of the materials that formed before the Nectaris basin). Part of the reason this scheme has worked so well globally is its tie to the formation of the Imbrium basin (Plate 3). This crater, at over 1,100 km diameter, is one of the largest, as well as the youngest, basins, and traces of its deposits can be found thousands of kilometers from the basin rim. Thus we are able to determine, at great distances from the center of the near side, whether a surface feature formed before (pre-Imbrian) or after (post-Imbrium) the Imbrium basin.

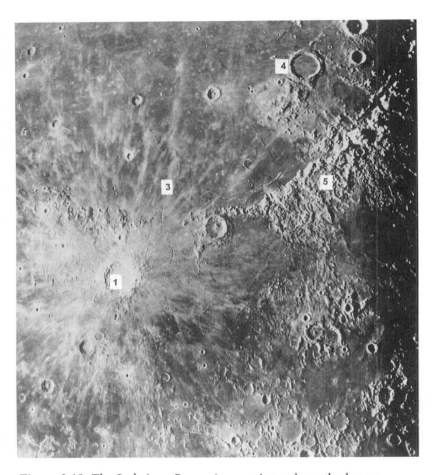

Figure 2.19. The Imbrium-Copernicus region, where the lunar relative age system was first worked out. Copernicus (1) overlies all other units. Eratosthenes (2) is postmare but pre-Copernicus. The maria (3) underlie these two craters but fill Archimedes (4) and the Imbrium basin (5). Archimedes (4) occurs within the Imbrium basin (5) and is therefore younger. Thus from this single photograph we can deduce the relative age of each of these units: Copernicus (youngest) to Imbrium basin (oldest).

Geological mapping tells us the *relative ages* of surface features, not their absolute age in years. For example, because of the high density of impact craters on the Moon, we knew before the Apollo missions that the vast bulk of lunar geological activity was Imbrian age and older. What we did not know was when the Imbrian Period started and when it ended. To add absolute ages to the relative timescale devised from geological mapping, we needed samples of units of regional significance. This is a relatively straightforward task for landing sites in the maria because the ages of most of the pieces of volcanic lava returned by the missions are presumed to reflect the age of the local bedrock, which in this case is made up of lava flows in the maria. Thus, the first Apollo landing told us that most of the maria dated from around 3.7 billion years ago. The ages of the craters and basins of the highlands would therefore be even older, a supposition later confirmed by landings in the highlands.

Things become a little dicey at the highland landing sites. From these later landing sites we received a bewildering variety of complex, mixed, remelted, and crushed rocks. When we date these samples, we may obtain extremely precise crystallization ages. But just what are we dating? Because these samples are obtained from the regolith at sites where the geological context is unclear, we are not sure exactly where they come from. For example, several different highland breccias were sampled at the *Apollo 17* landing site, near the edge of the large Serenitatis impact basin. Do these samples represent the ejecta from the impact that created the Serenitatis basin? Are they part of a basin "melt sheet"? Do the samples come from multiple impact events, some of them from craters and some from basins? The Imbrium basin is younger than the Serenitatis basin; is there Imbrium ejecta at this site? If so, which rocks come from that event?

I mention these difficulties because interpreting the absolute ages of surface features is a difficult and ongoing process. The absolute ages (in years) often seen alongside tables of relative age are not statements of fact; they are statements of interpretation. Even the age of so fundamental an event in the history of the Moon as the Imbrium impact, an event potentially sampled at every Apollo site, is both debated and debatable. The same prob-

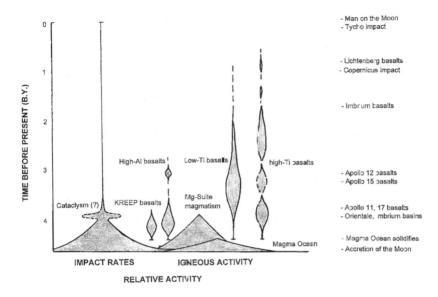

Figure 2.20. Diagram that shows the relative amounts of geological activity on the Moon as a function of time. Note that almost all activity on the Moon is very ancient, with most activity being confined to a period of time before 3 billion years ago.

lems attend the youngest end of the timescale. Our estimate of the time span of approximately 1 billion years for the Copernican System is inferred from an interpretation of the argon degassing age of a single sample of *Apollo 12* soil—soil that may or may not be part of the Copernicus ray in this area. Fortunately, the definition of the base of the Copernican System is approximate anyway, so approximate absolute ages suffice in this case. But a consequence of such vagueness is that we do not know the absolute age in years of the youngest lava flows, events crucial to understanding the thermal evolution of the Moon.

Our best estimate of the times, durations, and major events in the geological history of the Moon is summarized in Figure 2.20. Even given the cautions listed above, we possess a remarkably detailed knowledge of when events occurred on the Moon and the relative rates of geological activity with time. Clearly, the Moon is an ancient world. More than 99 percent of its surface area predates 2 billion years ago, whereas on the active and ever-changing Earth, areas older than 2 billion years occupy less than

5 percent of its surface. Our picture of lunar evolution, built up painstakingly over 30 years of mapping, sampling, remapping, and continuous reevaluation, is one of a small planet, intensely active for its first 600 million years, declining slowly over the next 500 million, and nearly quiescent for the remainder of its history, a silence punctuated only by the rare impact, causing the Moon to slowly oscillate like a bell.

Thus the story of the Moon is the story of the early solar system, a time when worlds collided, globes melted, planetary crusts shattered under the onslaught of a barrage of impacting debris, and dull-red, glowing lava flows coated the surfaces of lifeless worlds. The story is one of a truly alien landscape—and it was all reconstructed from the simple observations of a few carefully chosen features.

 Chapter 3

The Exploration of the Moon

The United States went to the Moon not to advance science but to demonstrate to the world that its system of values was superior to that of its Communist competitor. Likewise, the country stopped going to the Moon not because it had scientific reasons but because the political objectives responsible for the creation of the Apollo program had been satisfied. Although the story has been well told elsewhere (see the Bibliography), I want to briefly review the account of the exploration of the Moon because knowing something about how the task was approached helps us to appreciate more fully the nature of our current understanding of the Moon and its history.

A Bold Challenge and a Tough Problem

The launch of the Soviet satellite *Sputnik* in October 1957 sent shock waves through the very fiber of American society. How could a nation that we considered our technical inferior produce the world's first Earth-orbiting satellite? The ability of the Soviets to fly *Sputnik* presented a horrible implication: If they could launch a satellite, they could surely lob a nuclear warhead directly into the heart of the United States. Such a technical challenge could not go unanswered, and with the creation of the National Aeronautics and Space Administration (NASA) by Congress in 1958, the United States began its long, sputtering climb into the space age.

Initial American efforts were not auspicious. Failure and delay characterized the early space program, including spectacular fireballs as rockets exploded on the launchpad and during liftoffs. With the successful launch of *Explorer 1* into Earth orbit by Wernher von Braun and his colleagues in January 1959, the nation was finally on track. Over the next several years the United States seemed to be catching up to the Soviets as it orbited many satellites and prepared to send men into the unknown, following the trail blazed by robotic precursors. But once again the Soviets struck first, launching Yuri Gagarin into Earth orbit in April 1961, a full month before the Mercury program's first suborbital flight, by Alan Shepard in the spacecraft *Freedom 7* in May. This success of the Soviets, coupled with the rather spectacular failure of the invasion of Cuba by a CIA-trained Cuban army in exile, caused the new American president, John F. Kennedy, to search desperately for a field on which the Americans could challenge the Soviets. After due consideration, Kennedy decided to set a decade-long goal of landing a man on the Moon and returning him safely to Earth.

The goal of a manned landing on the Moon provided a suitable challenge—a task believed to be one that the United States could win. A look at the problem was sobering, however. A trip to the Moon and back would traverse over a million kilometers of space; at the time, the U.S. distance record with man was about 500 km. The trip would last at least a week; Shepard had been in space for about 15 minutes. A trip to the Moon would require a rocket with several million pounds of thrust, capable of carrying at least 100–300 tons into low Earth orbit; Shepard's *Redstone* booster rocket for *Freedom 7* developed about 80,000 pounds of thrust, and his Mercury spacecraft weighed about 1 ton. To advocate a trip to the Moon by a nation whose total experience in manned spaceflight was the flight of *Freedom 7* was audacity indeed!

The exact way that the United States would go to the Moon remained a contentious issue. The principal competing ideas in the "mode decision" (as it was called) revolved around space rendezvous, in which two spacecraft would meet in space to exchange people, cargo, or fuel. The question was, should the lunar craft be assembled for flight in Earth orbit, refueling from an orbiting tanker (a mode called *Earth-orbit rendezvous*), or should it be launched all at once, sending down a very small lander from lunar orbit and then returning a man back to the spacecraft orbiting the Moon (a mode called *lunar-orbit rendez-vous*)? Although lunar-orbit rendezvous seems natural to us to-day, in 1961 no one had ever achieved *any* type of rendezvous in space, let alone conducted one a quarter of a million miles away from Earth. What finally tipped the balance in favor of this risky technique was a study of the launch booster requirements for each mode; an Earth-orbit rendezvous would require a "super-booster" of over 14 million pounds of thrust. Although such a booster was designed in a preliminary manner, it was decided that our best chances were with the lunar-orbit rendezvous tech-nique, a mode that would require "only" a booster that could develop about 7.5 million pounds of thrust, a vehicle later called *Saturn 5* (Plate 4).

The fact that the United States was going to the Moon did not guarantee that we would explore it scientifically. However, some information about our destination was needed to ensure a safe voyage and landing. We needed to learn how to control space-craft at lunar distances, how to maintain an orbit around the Moon, and how to land and operate safely where we did not know the surface conditions. Such an abundance of ignorance virtually ensured that we would be undertaking precursor mis-sions, missions that would not only blaze the trail but also, invariably, advance our understanding of the Moon and its envi-ronment in major ways.

As described in Chapter 1, Eugene Shoemaker had antici-pated lunar voyages and had already been studying the Moon to prepare for this upcoming golden opportunity. Because the ba-sic geological framework of the Moon had been comprehended (see Chapter 2) and because its near-side topography and geol-ogy were being mapped and its environment characterized, it was a relatively straightforward task to devise an exploration plan that would logically address and answer the key un-knowns. We had to understand the surface layer, both to ensure a safe Apollo landing and to comprehend the geology of the returned samples. To find large smooth areas, we needed de-tailed maps of the potential landing sites on the Moon. We had to map the gravity field of the Moon, to ensure that the Command-Service Module (CSM) spacecraft would remain safely in precisely known orbits and to guarantee that the Lunar Module (LM) spacecraft could make a pinpoint landing and later rendezvous with the orbiting command ship. Such were the knowledge requirements for a voyage to the Moon; how were these requirements satisfied?

The Robot Precursors: Ranger, Surveyor, and Lunar Orbiter

Three principal flight projects added to our pre-Apollo understanding of the Moon (Fig. 3.1), and they still provide data of scientific value today. Collectively, they showed that the surface, though dusty, did not contain deep pools of quicksand-like dust waiting to swallow unsuspecting spacecraft. Smooth, boulder-free areas in the maria were identified and mapped. Nearly the entire surface of the Moon was photographed at resolutions 10 times better than could be obtained from Earth, allowing us to extend our geological mapping to the entire globe and illustrating the nature of the terrain types and their implications for lunar history. We even made the first chemical analyses of the surface, confirming the volcanic origin of the maria and finding something unusual and unexpected in the highlands.

The Soviets had beat us to the punch in photographing the Moon from up close when its *Luna 3* spacecraft succeeded in returning pictures of the far side in 1959. The U.S. Ranger program started in late 1959 and originally was a principally scientific mission, designed to obtain close-up views of the surface. Ranger was a hard lander and would destroy itself on impacting the Moon at near escape velocity (3.5 km/sec). Later missions were to have carried a crushable, balsa-wood ball that would allow the safe delivery of instruments to the lunar surface,

Figure 3.1 (opposite): The three principal robotic precursor missions that paved the way for the manned Apollo missions to follow. At top, the crash-landed Ranger spacecraft, designed to take pictures of increasing detail before impact on the Moon. At middle, the soft-landing Surveyor series, which certified that the surface could support machines and people and told us much about the processes of surface evolution. At bottom, the Lunar Orbiter spacecraft, which mapped the whole Moon and took extremely detailed pictures of proposed Apollo landing sites.

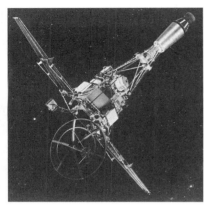

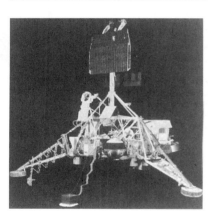

including a seismometer to measure *moonquakes*. These enhancements were dropped when the high-gear schedule of Apollo demanded immediate results. Such results were not instantly forthcoming from Ranger. Several Ranger flights either blew up on launch, missed the Moon completely, or silently crashed into the lunar surface without transmitting a single picture. Finally, on July 31, 1964, the *Ranger 7* spacecraft returned a spectacular series of close-up photographs of a portion of Oceanus Procellarum (Fig. 3.1), the largest expanse of maria on the Moon, each image showing a smaller and smaller area at greater resolutions than ever before.

From the *Ranger 7* mission, we discovered that the ubiquitous craters on the surface continue downward in size to the limits of resolution (even occurring on the surfaces of the returned rock samples). The Ranger photographs allowed us to decipher the nature and dynamics of the ground-up, powdery surface layer that everywhere covers the Moon. From the numbers of small craters, we learned about the effects of bombardment of the surface by micrometeorites and discovered the concept of crater *equilibrium*, in which the rate of crater production by impact equals the rate of crater destruction by erosion. All lunar surfaces are in cratering equilibrium at some diameter. The larger this equilibrium crater diameter, the older the surface. The oldest surfaces of the highlands have equilibrium crater diameters of tens of kilometers, indicating that they are saturated with very large impact craters and that the crust is crushed and broken by impacts for depths of many kilometers.

The next mission, *Ranger 8*, was sent in early 1965 to the western edge of Mare Tranquillitatis. Once again, we saw the crater-upon-crater texture typical of the lunar surface at close ranges. The *Ranger 8* mission also photographed two unusual craters, Ritter and Sabine, thought at the time possibly to be large volcanic craters. With earlier missions having scouted two different regions of the maria, the last Ranger mission (*Ranger 9*, March 1965) was sent to an area in the highlands, the spectacular ancient crater Alphonsus (Fig. 5.4), on the eastern edge of Mare Nubium. Alphonsus represents a class of feature called a *floor-fractured crater;* cracks found on the floor of the crater, in addition to small, dark-rimmed craters that might be cinder cones, are thought to be manifestations of volcanism. Indeed, Alphonsus had long been one of the sites of the famous *lunar transient phenomena*—reddish, glowing clouds had been reported emanating from the crater. *Ranger 9* saw no evidence for gas venting from the Moon but did return spectacular images of Alphonsus at ever higher resolutions. Another remarkable facet of the *Ranger 9* mission was the return of its pictures via real-time, live television from the Moon, watched by fascinated people across the nation (including me, age 12) as the probe struck the Moon at high velocity.

As Ranger finished giving us our first close look at the Moon, we prepared to soft-land and touch its surface for the first time. Surveyor was originally a spacecraft with orbiter and lander elements, but faced with a choice imposed by delays, the program dropped the orbiter portion and concentrated work on the lander to best support Apollo (Fig. 3.1). Initial Soviet attempts to soft-land on the Moon failed repeatedly, raising questions about the surface conditions. One lurid and widely reported model suggested that the maria were gigantic bowls of dust, cauldrons that would act like quicksand to swallow up any equipment landed on the Moon. Geologists mapping the Moon with telescopic pictures considered such models nonsense but would have to wait for the first successful landing to put such fears permanently to rest.

After much effort, the Soviets beat the United States to the punch once again (for the last time, as it later turned out) and succeeded at a soft landing on the Moon in February 1966 with their *Luna 9* spacecraft. Television pictures showed a surface similar to hard-packed sand covered with a thin layer of dust. The first American landing, *Surveyor 1* in early June 1966, returned hundreds of detailed pictures of the surface, showing us the Moon as it would appear to an astronaut standing on the surface. The Surveyor pictures showed the ground-up surface layer, documenting that it was strong enough to support the weight of the people and machines that would soon be visiting. We could see in the Surveyor pictures evidence for the mixing and crushing of the bedrock into the rock-laden, dusty layer of dirt that makes up the surface.

Five Surveyor spacecraft (*1, 3, 5, 6,* and *7*) successfully landed on the Moon; radio contact with *Surveyors 2* and *4* was lost shortly before landing, and they are presumed to have crashed. *Surveyor 3* carried a trenching tool, designed to dig into the surface and study its properties and strength at depth. This trenching scoop and a television camera were returned to the Earth nearly three years later by the *Apollo 12* astronauts and showed that hardware could withstand long exposure to the lunar environment (the camera is now on display at the Smithsonian Air and Space Museum). *Surveyor 5* carried the first experiment designed to measure the Moon's composition, an instrument for determining the chemistry of the surface. From these data, we found that the maria were rich in magnesium and poor in aluminum, results consistent with a composition of basalt, a very common type of lava on Earth. Having landed *Surveyor 6* at another mare site (Sinus Medii), the program sent the last mission, *Surveyor 7*, to one of the most spectacular locales on the Moon: the rough, hazardous rim of the crater Tycho (Fig. 2.4), deep in the southern highlands. *Surveyor 7* beat the odds by safely landing at Tycho in January 1968.

Although *Surveyor 7* returned fascinating views of the rim of a complex crater (Fig. 3.1), its most significant experiment was the first determination of the chemistry of the highlands. The data showed a surface relatively rich in aluminum and depleted in magnesium, the reversal of the trends seen in the data from the maria. The team analyzing this information suggested that unusual rock types, including one called *anorthosite*, might be the main components of the highlands. Anorthosite and related rocks are made up mostly of a single type of mineral, plagioclase (a calcium-and aluminum-rich silicate). If this supposition was correct, it would have significant implications about the history of the Moon. This chemical composition was anything but primitive, contradicting the concept that the Moon was a cold, undifferentiated object, as Harold Urey believed (see Chapter 1).

The last major task for the robotic precursors to Apollo was the making of detailed maps of the Moon. Such maps were needed to choose and to certify safe landing sites and to aid the astronauts in exploring the new world. If the missions were completely successful, a significant by-product of this effort would be the global, scientific mapping of the Moon. Five of the unimaginatively named Lunar Orbiter spacecraft (Fig. 3.1) flew between August 1966 and August 1967, and each one was an overwhelming success. Unlike the design of the other precursor missions, the plans for the Lunar Orbiter camera were based on the design of classified, espionage spacecraft, intended to photograph features at high resolution from space. The first three Orbiter spacecraft were placed in near-equatorial orbits, similar to those to be used by the upcoming Apollo missions, and returned dozens of very high resolution pictures of the proposed landing sites (Fig. 3.1). Features as small as one-half of a meter in size were recognized, classified, and mapped. Detailed mosaics were made to aid in plotting the landing paths of the Apollo LM spacecraft as well as to depict the boulders, lineaments, and geological features that the astronauts might explore.

With the scouting of the landing sites accomplished, the last two Orbiter spacecraft were sent into near-polar orbits so that the entire surface would come into the view of their cameras. *Lunar Orbiter 4* mapped the entire near side at a resolution 10 times better than the very best views from Earth. It also took detailed and spectacular images of the Orientale impact basin (Fig. 2.8), images from which we first learned details about the process of large impact and the creation of the strange landforms in the highlands. *Lunar Orbiter 5* made detailed, very high resolution mosaics of sites of high scientific interest, including the fresh craters Copernicus, Tycho, and Aristarchus and volcanic regions such as the Marius Hills, the Rima Bode area, and Hadley Rille, a future Apollo landing site. The pictures returned by the Lunar Orbiter series not only paved the way for the Apollo missions but also gave us images of the Moon that are still used extensively by scientists today (as the reader will note through the use of these photographs for many of the illustrations of this book).

The Lunar Orbiter missions also revealed an unexpected hazard for voyagers to the Moon. The orbits of the spacecraft changed with time because subsurface zones of high-density material would tug at the spacecraft, gradually pulling them toward the Moon. The effect of these zones could be sudden and catastrophic: A small satellite released into lunar orbit by the *Apollo 16* mission lasted only two weeks before crashing into the

Moon. These regions, called *mascons* for "mass concentrations," are associated with the large, circular maria and were our first unintended "probe" of subsurface conditions—the Orbiter missions made our first *gravity map* of the Moon. Several models for the formation of the mascons were proposed. The two principal ones involved a thick fill of the basins by flows of high-density lava or the uplift of dense rocks from the mantle by the unloading of the crust during excavation of an impact basin. We now think that the uplift of mantle rocks is the dominant cause of the mascons, but some contribution from lava flooding is probable. More important for lunar exploration, the mascons were a potential hazard for the upcoming Apollo missions, and their effects on the orbits of spacecraft around the Moon had to be comprehended before the missions could be successful.

These precursor missions successfully paved the way for the Apollo missions. We had achieved a broad understanding of lunar history and processes through geological mapping at a variety of scales. We had carefully measured the physical properties of the surface and assured ourselves that it would not swallow up the people and equipment we would soon be sending. Almost as important, we had acquired real operational experience with the flying and operating of spacecraft at lunar distances. Coupled with the human spaceflight experience of rendezvous, docking, and *extravehicular activity* (EVA, or "spacewalking") in Earth orbit during the Gemini program, we were ready to send men to the Moon.

Man Orbits the Moon: *Apollos 8 and 10*

The first humans to look at the Moon close-up were the crew of *Apollo 8*—Frank Borman, Jim Lovell, and Bill Anders—who orbited the Moon in December 1968. *Apollo 8* was only the second manned Apollo flight (*Apollo 7* had conducted an Earth-orbiting mission in October 1968) and the first manned flight of the mighty *Saturn 5* rocket (Plate 4). Sending this mission all the way to the Moon after one Earth orbital flight of the Apollo spacecraft was a bold step. *Apollo 8* made ten orbits of the Moon, taking photographs of the far side and of the far eastern Apollo landing site, finding it to be unexpectedly rough, and making detailed visual observations of the surface. As had been feared from the Lunar Orbiter data, the spacecraft orbit was indeed disturbed by the presence of the high-density, subsurface mascons, and this hazard would have to be understood and dealt with during the upcoming landing attempts. One of the most significant emotional effects of the *Apollo 8* mission was its famous photograph of a beautiful, blue-green Earth appearing to rise slowly above the stark, gray "wasteland" of the Moon.

After the *Apollo 9* mission tested the LM in Earth orbit in March 1969, the *Apollo 10* mission in May 1969 conducted a full-up dress rehearsal for the lunar landing. Both the CSM (flown by John Young) and the LM (flown by Tom Stafford and Gene Cernan) orbited the Moon, the LM descending to within 15 km of the surface. *Apollo 10* made detailed observations of "Apollo site 2" in Mare Tranquillitatis and confirmed that its smooth appearance on the Lunar Orbiter mosaics was real and that the site seemed to be an appropriate place to attempt a landing. A nearly complete moon-landing mission profile was flown, with the LM firing both its descent and its ascent engines in a test of the Apollo spacecraft in real lunar flight. After several separate orbits, the LM returned to the CSM to rendezvous and dock, just as would the lander returning from the surface. Another day was spent orbiting the Moon and taking additional photographs of the area near Sinus Medii, where *Surveyor 6* had landed just two years earlier. Both the *Apollo 8* and the *Apollo 10* orbital missions paved the way for the grand act to follow: the completion of the decade-long challenge laid down a mere eight years previously.

Man on the Moon: *Apollo 11*

The first lunar landing, on July 20, 1969, was a real cliffhanger—almost literally. As the *Apollo 11* LM *Eagle* swooped in for its landing, Commander Neil Armstrong noted that his spacecraft was headed for the center of a crater, several hundred meters across and strewn with blocks, some the size of small automobiles. Armstrong took manual control, carefully flying *Eagle* around the obstacle, and set the craft on the Moon with less than a few seconds of hovering fuel left in its tanks. Neither Armstrong nor his LM pilot, Buzz Aldrin, recognized exactly where they were on the Moon. Once again, the mascons were to blame; they had pulled *Eagle* downrange, off course by several kilometers (in fact, completely outside of the designated landing area). Even the orbiting CSM pilot, Mike Collins, could not see the LM on the surface; it was not until the astronauts returned home and the onboard film was reviewed that their landing site was located precisely. At the time of the landing the astronauts knew only that they had successfully pulled off the greatest feat in the history of flying since Orville Wright's 10 seconds of immortality. A few hours later Armstrong took his "one giant leap" and became the first human to walk on another world.

In addition to fulfilling the dreams of millennia (Fig. 3.2), the first landing on the Moon accomplished quite a bit of science as well. The astronauts set out a small seismometer, which documented that the Moon is extremely quiet and that moonquakes are small and rare. They also deployed a laser reflector, with which we could measure, to within a few centimeters, the distance between Earth and the Moon. Such precision measurements would allow us to carefully track the Moon's orbital motion and, in addition, the drift of the continents on Earth. The astronauts collected about 40 kg of rock and soil samples from the immediate vicinity of the LM, and the return of this material to Earth answered some of the most important scientific questions that had accumulated over the years.

The lunar samples were extremely dry; no evidence was found for any water whatsoever, with a complete absence of water-bearing minerals in the rocks. The rocks were samples of either basalt, a common lava type on Earth, or breccia, a rock made up of many fragments of older rock and minerals. The basalts from the *Apollo 11* site contained relatively high concentrations of titanium, an unexpected enrichment. It was found that the lunar soil is composed of ground-up lava bedrock and includes abundant tiny fragments of glass, created by the shock melting of small mineral grains during high-velocity impact. The ages of the rocks, determined by measuring the amounts of radioactive isotopes in the materials, were found to be very great; the lavas of Tranquillity Base flowed 3.7 billion years ago, long before virtually all of the surface rocks of Earth had been created. Strangely, the soil (which was created from and lay on

Figure 3.2. An age-old dream achieved. *Apollo 11* Lunar Module pilot, Buzz Aldrin, on the Moon, July 20, 1969. The mission commander, Neil Armstrong, is visible, reflected in Aldrin's faceplate.

top of the 3.7-billion-year-old rocks) appeared to be even *older*, dating 4.6 billion years, back to the age of the Moon itself. It took another mission to the Moon to explain this puzzling fact: The soil is relatively enriched in a radiogenic isotope of lead, resulting in ages that are only apparently older.

The results of the *Apollo 11* mission placed some broad constraints on lunar history. The maria are made of volcanic rock and are very old. Because basalt lava forms by partial melting of a certain type of rock, the *Apollo 11* basalts showed that the interior of the Moon was not primitive in composition but had been created in an earlier melting episode. The surface layer (regolith) is made of ground-up bedrock, partly crushed into powder and partly fused by impact melting. One finding yielded a major insight into lunar evolution: Tiny, white fragments found in the soil are clearly different from the local bedrock. It was postulated that these fragments are pieces of the highlands, thrown to the *Apollo 11* site by distant impacts. This supposition was supported by the chemical analysisthat the *Surveyor 7* spacecraft made of Tycho ejecta, which showed the unusual aluminous composition described earlier. If the highlands were really made of this unusual rock type, anorthosite, the early Moon may have been nearly completely molten, an astonishing idea for a planet as small as the Moon. This concept, called the *magma ocean*, was reinforced by subsequent mission results and will be described in Chapter 6.

Deepening Mysteries: *Apollos 12* and *14*

The second landing, in November 1969, was considerably more lighthearted than the first. Leaving their colleague, Dick Gordon, in lunar orbit, astronauts Pete Conrad and Alan Bean demonstrated a new technique for pinpoint landing. Correcting for the effects of the dreaded mascons, they landed only a few tens of meters from where the *Surveyor 3* spacecraft had landed in eastern Oceanus Procellarum. This landing site, like that of *Apollo 11*, was in the maria and featured deposits that were slightly less cratered (and therefore younger) than those sampled at Tranquillity Base. We did not know how much younger the flows were. This mission featured two moonwalks, each over three hours long, and the emplacement of a nuclear-powered geophysical network station (Fig. 3.3), the ALSEP (Apollo Lunar Surface Experiment Package—the Apollo engineers loved acronyms). The crew collected more rock samples than the *Apollo 11* crew and visited the rims of several impact craters at their landing site. The *Surveyor 3* spacecraft (Fig. 3.1) was visited and sampled to assess the effects of long-term exposure on the surface of the Moon (the effects were not very noticeable—a few microcraters).

Figure 3.3. An *Apollo 12* astronaut adjusting the antenna of the central station of the Apollo Lunar Surface Experiment Package (ALSEP), the network of geophysical stations set up by five of the six Apollo landing crews. The three-prong instrument in the foreground measures the magnetic field.

The *Apollo 12* basalts showed the same extreme age and lack of water as the other lavas but with some important differences. First, the basalts of the *Apollo 12* site were lower in titanium than those from Tranquillity Base, demonstrating that different regions of the interior had melted to make the two types of lava. Second, these lavas, themselves representing several different lava flows, were "only" 3.1 billion years old, almost 500 million years younger than the lavas from the other site. These results showed that the maria did not all result from a single, massive volcanic eruption but represented a complex series of lava flows poured out over at least a half a billion years. Rare fragments of highland rocks from the *Apollo 12* site included some that were quite different from those found at the *Apollo 11* site, demonstrating that the highlands similarly varied from place to place. An impact breccia from this site is an extremely complex mixture of unusual rock types, foreshadowing similarly complex breccias to be returned by future missions to the highlands. A strange enrichment in certain elements—including potassium, phosphorous, and some radioactive elements—was first recognized in soils and rocks from this site. This material, given the name KREEP, is an important clue to the origin of the crust.

The *Apollo 13* mission in April 1970 was to be sent to the Fra Mauro highlands, just east of the *Apollo 12* landing site (Fig. 3.4). Unfortunately, an oxygen tank in the CSM of the spacecraft exploded on its way to the Moon, and after a truly heroic emergency effort, including the use of the LM as a "lifeboat" to sup-

Figure 3.4. The crater Fra Mauro (center), which is overlaid by a rough blanket of debris called the Fra Mauro Formation. The *Apollo 14* landing site (arrow) was sent here to sample rocks thrown out of the Imbrium impact basin.

port the crew, Jim Lovell, Fred Haise, and Jack Swigert returned to Earth safely after looping around the Moon. When lunar spaceflight was resumed in January 1971, the *Apollo 14* mission was redirected to the unvisited site. Fra Mauro was considered to be important because the site was on the regional blanket of debris (Fig. 3.4) thought to be thrown out of the Moon by the impact that created the huge Imbrium basin, a crater over 1,000 km in diameter. Alan Shepard, the nation's first spaceman, and Edgar Mitchell directed their LM *Antares* to a pinpoint landing on the Fra Mauro ejecta blanket. The CSM was piloted by Stu Roosa, who conducted an extensive program of observations from orbit. During two moonwalks, Shepard and Mitchell set up another ALSEP station and fired small explosive charges to "profile" the subsurface with seismic lines (much as is done on Earth during oil prospecting). After trekking up the slopes of a hill 3 km distant and 100 m high to explore the ejecta from Cone crater (1 km diameter), which excavated the Fra Mauro debris blanket, the crew brought up deep rocks for our examination and collection.

The astronauts became disoriented and lost during their moonwalks. The crystal clarity of pure vacuum and the lack of recognizable landmarks confuse the mind and make it very difficult to judge distances on the Moon. Features that appear nearby may be many kilometers away. The smooth, rolling nature of the highlands means that even when one stands on high ground, areas that are physically close may be unseen while distant craters may be clearly visible. Apparently, the astronauts literally walked right by the rim of Cone crater without seeing it. Even so, Shepard and Mitchell did succeed in returning samples of the ejecta of Cone crater, and these rocks both confused and enlightened scientists. The problem of navigating on the lunar surface was solved on the next mission by letting a computer on the Lunar Roving Vehicle (LRV), or *rover*, keep track of where the astronauts were at any given time.

The materials returned by the *Apollo 14* mission are some of the most complex rocks in the sample collection. They are all breccias, complex mixtures of older rocks, including breccias containing breccias from previous events. Nearly all are enriched in the strange KREEP chemical component first identified at the *Apollo*

12 site, and in bulk composition they are considerably different from what was expected. It had been thought that highland rocks would be extremely rich in aluminum, made of anorthosite, as found at the *Apollo 11* site. In fact the bulk composition of the *Apollo 14* breccias is basaltic, not as iron-rich as the mare samples but still much less aluminous than anorthosite. This unusual composition told us that the highlands are composed of different provinces, possibly reflecting different geological histories and evolution. In this case the composition of the Fra Mauro breccias is related to their origin as ejected debris from the giant Imbrium impact basin.

The breccias from this site also showed us that mare volcanism began very early in lunar history. Basalts that were found as small fragments embedded in these breccias are 4.2 billion years old, nearly as old as the crust itself. From this site, we also recognized another important rock type, the *impact-melt breccia*, a rock that strongly resembles volcanic lava but that was created from the intense shock pressures of impact rather than from a volcanic eruption. Impact melts are important components of the highlands because they are the impact products most suitable for radiometric dating. Thus they can tell us much about the geological history of the Moon, provided they can be related to the crater that formed them.

The Great Explorations: *Apollo 15, 16,* and *17*

The last three Apollo missions, in 1971 and 1972, introduced a new and exciting scale of exploration, a scale not surpassed (or even equaled) today. Each mission consisted of an upgraded, expanded spacecraft, allowing more experiments and more sophisticated equipment to be carried to the Moon. The orbiting CSM carried a special package of cameras and sensors to study the Moon from orbit. Each mission carried an electric cart, the rover (Fig. 3.5), to the surface. This innovation was not a gimmick but was a valuable exploration tool that permitted the astronauts to venture farther from the LM (by navigating across the surface) and to stay longer at scientifically important sites (by permitting the crew to rest and conserve their air and water while traveling to distant sites). In addition a drill and coring rig

Figure 3.5. An image from the *Apollo 17* mission. On the last three missions to the Moon, an electric cart, the Lunar Roving Vehicle (LRV), allowed the astronauts to travel farther and explore more territory. The large dish at right allowed the LRV to transmit television pictures, even when the rover was far away from the Lunar Module.

allowed deep samples of the regolith to be obtained. All three missions deployed an advanced ALSEP package, creating a long-lived geophysical network that would continue to send data from the Moon back to Earth until the network was turned off six years later. Each LM could stay on the Moon for up to 72 hours, almost doubling the exploration time. This new exploratory capability was exploited by sending these last three missions to complex, multiple-objective landing sites.

Figure 3.6. The magnificent *Apollo 15* Hadley-Apennine landing site (arrow). The site was picked to allow the astronauts to examine both the snakelike Hadley Rille, an ancient lava channel, and the towering Apennine Mountains (right), which make up the rim of the Imbrium impact basin.

The *Apollo 15* mission was sent to the rim of the Imbrium basin at the spectacularly beautiful Hadley-Apennine landing site in July 1971. The huge chasm of the sinuous Hadley Rille (over 2 km wide and 900 m deep) winds across the mare plain, surrounded by one of the steepest, highest (7 km) mountain ranges on the Moon (Fig. 3.6). It provided viewers on Earth with their most memorable lunar scenes. This site was the first multiple-objective site, with the crew being able to sample and explore both mare terrain and the highlands bordering the Imbrium basin. The three astronauts—Dave Scott, Jim Irwin, and Al Worden—received extensive training in geology, hard work that paid off magnificently as they explored this corner of the Moon for three days. Time outside the LM more than doubled, and traverse distance increased by a factor of five over the previous mission as the astronauts drove the rover across the dusty plains at Hadley.

Apollo 15 returned a variety of impact breccias from the highlands and mare basalts from the plains, but there were also a few surprises in the sample box. A transparent, emerald-green glass was discovered scattered about the site. Analysis showed that this glass is a form of ash deposit from a volcanic eruption over 3 billion years ago. Small fragments of lava rock with aluminum-rich composition were our first sample of "nonmare" or highland volcanism. Detailed photographs showed unusual benches in the mountains of the surrounding highlands (Fig. 3.7), possibly exposing layered rocks from the period of early bombardment. Similarly, Scott and Irwin visited the rim of Hadley Rille, which exposed layered rocks in its walls, mute testimony to a prolonged filling of the Imbrium basin by separate lava flows over a period of many years. The *Apollo 15* mission was the most extensive exploration of the Moon yet, a tribute to the scientists and engineers who were determined to make Apollo a genuine tool of exploration.

The *Apollo 16* mission, in April 1972, is often referred to as the only mission to the highlands, but this is incorrect: *Apollo 14* was also sent to a highland site (Fig. 3.4). *Apollo 16* was, however, the only Apollo mission whose site was distant from the maria; the landing site was located in the central highlands, near the ancient crater Descartes (Fig. 3.8). This mission is most renowned for having disproved its preflight predictions: Planners had believed that the Descartes site was composed of light-toned, highland volcanic rocks, including ancient ash flows and silica-rich dome volcanoes. The LM crew—veteran pilot John Young and newcomer Charles Duke—and the CSM pilot, Ken Mattingly, were given extensive training in volcanic terrains on Earth to prepare them for the exploration of Descartes. The skilled crew members were surprised during their moonwalks—

Figure 3.7. A telephoto view of the mountains at the *Apollo 15* landing site. The large ridge is called Silver Spur and may have resulted from the exposure by impact of ancient, deeply buried layers of rock. This cliff is over 600 m high.

where were all the volcanic rocks they expected? Instead almost every variety of impact breccia imaginable was found at every sampling site (Fig. 3.9).

Essentially two geological units were sampled during the *Apollo 16* mission: the wormy-textured Descartes mountains and the smooth, light-toned Cayley plains (Fig. 3.8). Each unit is made up of impact breccia, and scientists still debate whether there is any compositional difference between the two units. The rocks are made up of regolith breccias (found at all sites), fragmental breccias (made up of fragments of older rock), and impact-melt breccias (comparable to those returned previously by the *Apollo 14* and *15* missions). The melt rocks are of particular interest. These breccias have a composition distinct from the upper crust in this region; in fact, their composition more closely resembles

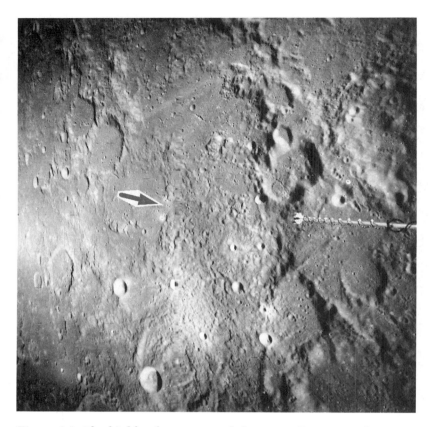

Figure 3.8. The highland area around the crater Descartes, the location of the *Apollo 16* landing site (arrow). Although ancient volcanic rocks were expected, the mission returned impact-processed rocks instead. This finding led to a major revision of our understanding of the Moon.

melts believed to be ejecta from the huge Imbrium basin, sampled earlier. How can this be, considering the great distance of the *Apollo 16* site from Imbrium? Perhaps this composition is common to many different basins, the Descartes site being rather close to the older Nectaris basin, about 300 km to the east. In any event the first Apollo mission to a "pure" highland site completely changed the way we look at the highlands. We now think that impact processes of various types, usually associated with basins, are responsible for the units that make the terrae look like a

Figure 3.9. *Apollo 16* Mission Commander John Young on the rim of North Ray crater, a large impact feature in the Descartes highlands. It is very difficult to judge distances on the Moon. The large boulder behind Young is farther away than it looks and is as big as a house.

patchwork quilt, and the role of volcanism in shaping the geology of the highlands is believed to be minor.

The final Apollo mission to the Moon was sent to the rim of the ancient Serenitatis basin, where mare lavas partly flood an ancient mountain valley. Dark mantle material coats the nearby hills; this unit was thought to be a young volcanic deposit, and it was believed that cinder cones might be found near the landing site. Astronauts Gene Cernan and Jack Schmitt (Schmitt was the first and, so far, the last professional geologist to explore the surface) landed in the Taurus-Littrow valley in December 1972; the orbiting CSM pilot was Ron Evans. The *Apollo 17* crew traveled the farthest (almost 30 km), explored the longest (over 25 hours), and collected the most samples (more than 120 kg) of all the Apollo missions. While exploring the Moon, the astronauts found and sampled giant boulders that had rolled down the mountains—a bright "landslide" triggered by the formation of the crater Tycho, over 2,200 km away—and beautiful orange soil that glistened in the bright sunlight of the surface (Plate 5). This spectacular flight was indeed a fitting finale to man's first round of lunar exploration.

As at the *Apollo 15* site, two major terrains were explored during the *Apollo 17* mission: the mare fill of the valley and the highlands of the surrounding Taurus Mountains. The mare lavas of the valley, basalts from many different flows, are very rich in titanium, similar to the lavas sampled by the first landing, *Apollo 11*. Photographs and remote-sensing data show that these high-titanium lavas may be continuous parts of the same series of lava flows in this region. The orange soil discovered at Shorty crater (Plate 5) turned out to be an unusual black-and-orange glass (Plate 6). Like the green glass from *Apollo 15*, these glasses are volcanic ash, the product of a huge fountain of liquid rock sprayed out onto the surface. However, this ash is old, not young as had been thought by the premission analysis, having erupted about 3.6 billion years ago, just after the eruption of the lava flows. Study of samples from the bright landslide across the valley (Plate 5) indicates that the crater Tycho formed 108 million years ago, providing an important time marker to the lunar geological column. The highlands, sampled at two different mountains (or *massifs*) at different ends of the valley, are made up of a complex mixture of rocks cooled slowly at depth and excavated from the deep crust by the giant impact that created the Serenitatis basin. Various impact-melt breccias were collected and found to be grossly similar to, but different in detail from, the melt breccias collected at the other Apollo sites. The most populous group of impact melts from the *Apollo 17* site may represent the melt sheet created during the impact that formed the Serenitatis basin, as is seen in the center of the Orientale basin (Fig. 2.9).

Figure 3.10. The Command-Service Modules of the Apollo spacecraft. The cylindrical module contained a bay of scientific sensors on the last three missions, allowing scientists to map the composition of the surface from orbit. The conical module (at bottom) is the only part of the Apollo spacecraft that returned to Earth.

These advanced Apollo missions carried sophisticated experiment packages in lunar orbit (Fig. 3.10). Two cameras and a laser altimeter measured the topography and shape of the Moon and took high-resolution stereo photographs, permitting detailed geological studies of different regions and processes. Sensors measuring the X-rays and gamma-rays permitted scientists to measure the chemical composition of the lunar surface. From these remotely sensed data, we first learned about regional provinces of different composition in the highlands. Other instruments detected the emission of gas from certain areas on the Moon, suggesting that the deep interior may still contain small pockets of volatile elements. On the *Apollo 15* and *16* flights, a small subsatellite was launched from the CSM, the first extraterrestrial launch of a satellite from another spacecraft. This subsatellite carried a magnetometer to measure the small magnetic field anomalies on the surface. The subsatellites also permitted the detailed gravity structure of the Moon to be mapped by tracking their radio signals. As mentioned earlier, the *Apollo 16* subsatellite lasted only two weeks before its orbit decayed because of the mascons and crashed into the Moon.

The end of the Apollo flights did not end the collection of lunar data. The sample collection, kept in a hurricane-proof vault in Houston, Texas, continues to be dissected, examined, and studied. The network of ALSEP stations on the Moon continued to send back data until it was decided to terminate the network in 1977 because of budgetary pressures. From the seismic experiment, we discovered that the Moon has an aluminum-rich crust about 60 km thick, beneath which is an iron- and magnesium-rich mantle. Probes designed to measure heat flow allowed us to estimate the amount of radioactive elements deep within the Moon and, from this, its bulk composition. We determined that the Moon has a composition similar to the mantle of Earth, and study of the isotopes of oxygen show that Earth and the Moon were made in the same part of the solar system. Both of these facts are significant constraints to models of lunar origin.

Measurements of the magnetic field of the Moon showed that local areas of the crust are magnetized, but the Moon does not possess a global magnetic field like that of Earth. Together with the relatively low bulk density of the Moon (about 3.3 g/cm^3, compared with 5.5 g/cm^3 for Earth), the lack of a global magnetic field suggests that the Moon has no large, liquid iron core, which generates Earth's global magnetic field by a process known as a core dynamo

The *Apollo 15, 16,* and *17* missions were outstanding successes by any objective measure. The Apollo program as a whole and these missions in particular form the cornerstone of our understanding of the Moon and its history. A nearly constant debate rages in the science community regarding the value of human spaceflight versus the unmanned robotic missions. The Apollo landings demonstrated that the difference, in capability and knowledge returned, between human exploration of the Moon and small robotic missions is comparable to the difference between a nuclear bomb and a firecracker. The manned Apollo missions revolutionized our understanding of the Moon and of planetary science in a way that the unmanned robotic precursors did not and could not. The Apollo missions are lasting testimony to the value of people in the exploration of the solar system.

The Russians Went Too: Soviet Robotic Lunar Landers

In retrospect it is reasonably clear that we were indeed in a race to the Moon with the Soviets in the decade of the sixties. Soviet leaders (especially Nikita Khrushchev but also his successors) considered space spectaculars to have enormous propaganda value, with each decisive "space first" demonstrating the superiority of "progressive" Soviet science and technology over the "decadent hedonism" of the capitalist West. Because of the catastrophic explosions of their giant booster rocket (the N-1) on at least two occasions, the Soviet manned landings on the Moon never took place.

Even after the race had been lost, the Soviets made a major effort to steal some of the thunder from the Apollo program. The most notable attempt was the flight of the mysterious *Luna 15* spacecraft in July 1969, at the same time that *Apollo 11* went to the Moon. *Luna 15* was an automated spacecraft that crashed into the surface while the *Apollo 11* crew was still orbiting the Moon. Soviet news releases clumsily issued reports that their "automated moon craft completed its historic mission," but at the time there was much speculation that *Luna 15* had been designed to land and to return a scoop of lunar soil to Earth before the *Apollo 11* astronauts could bring back their samples. After the *Apollo 11* crew returned successfully, the Soviets claimed that there had never been any race to the Moon with the Americans, a ludicrous claim then and now yet one believed and repeated by many in the credulous American media.

That the mission of *Luna 15* was indeed to return a sample is suggested by the flight of *Luna 16* in September 1970. This small spacecraft (Fig. 3.11) successfully landed on Mare Fecunditatis, on the eastern edge of the near side, and returned about 100 g of soil with an ingenious drill core that was wound into a ball-shaped return capsule. The soil from this site consists of mare

Figure 3.11. The Soviet *Luna 16* spacecraft, which returned a soil sample to Earth. Although conducted largely to steal some of the thunder from the Apollo program, these missions demonstrated that robotic return of planetary surface samples is technically feasible.

regolith, including several fragments of lava that were large enough to measure their ages. The *Luna 16* samples, mare basalts with relatively high aluminum content, erupted onto the surface 3.4 billion years ago. Abundant fragments of impact glass are also present, giving us clues to the existence of other, unsampled rock types on the Moon. The *Luna 20* mission landed on the Moon in February 1972 and was an identical copy of the *Luna 16* mission. It returned soil samples from the highlands surrounding the Crisium impact basin. The small samples are made up of tiny rock fragments of the highland crust, as at the *Apollo 16* site, and impact breccias. The final Soviet sample-return mission, *Luna 24* in August 1976, returned the largest sample to date: a 2-m core sample from the interior of Mare Crisium. These basalts are also a high aluminum variety but contain much less titanium than any Apollo sample (similar, very low titanium basalts were subsequently discovered in the soil from the *Apollo 17* landing site). The *Luna 24* basalts show that the lava flows in Mare Crisium erupted between 3.6 and 3.4 billion years ago.

The three Soviet samplers demonstrated that the robotic return of surface samples is a technically practicable tool to explore the Moon. Because of launch, flight-control, and landing constraints, these missions unfortunately were confined to landing sites on the eastern limb. It is highly desirable to be able to return samples from anywhere on the Moon. The Luna missions also show that there is not only a quantitative difference between human and robotic missions but a qualitative difference as well. We learned more about the Moon from any single Apollo mission than we did from the totality of the three Luna missions. This difference not only is related to the small mass of the returned sample from the Luna missions but also is caused by the geologically guided sampling that people can do. We understand the context of the Apollo samples much better than we do that of the Luna mission samples.

A variety of flybys, orbiters, hard landers, and rovers was also sent to the Moon by the Soviet Union. The two *Lunakhod* spacecraft were small rovers, remotely controlled from Earth. They had crude instruments and returned mostly television pictures and some data on the physical properties of the soil. However, they demonstrated that remote control of machines on the Moon is feasible. Future robotic sample-return missions should include the ability to operate the spacecraft remotely (teleoperation) so that the most significant samples can be obtained (see Chapter 10). Because the surface of the Moon is complex and varies from place to place, the ability to rove across its surface will be highly beneficial in future sample-return missions.

The Soviet lunar program, though not successful in its political objective to technically embarrass the United States, nevertheless achieved some significant scientific accomplishments that add to and enhance our understanding of the Moon. These small missions also foreshadowed the rich possibilities of small robotic spacecraft as tools for the exploration of the solar system. We will examine a variety of possible missions and their relative strengths and weaknesses when we consider future strategies for exploring the Moon (see Chapter 10).

Chapter 4

A Fall of Moondust
The Regolith

We have known for a long time that the surface of the Moon is covered with fine dust. If its surface were bare rock, we would see a bright reflection (a *specular* reflection) at the point on the Moon directly under the Sun (subsolar point) as it rotates on its axis. Such an effect is similar to the bright glare one sees when looking toward the Sun in late afternoon on a lake, ocean, or smooth body of water. Bare rock would also show this effect, although not as mirror-like as water because a rock surface would be much rougher at fine scales.

The surface of the Moon does not display this type of bright specular reflection. In fact by carefully studying the exact way light is reflected from the surface, scientists determined that the Moon was covered everywhere by dust—very fine dust. What was not known in the pre-Apollo days was how deep such a dust layer might be. Some scientists thought that it might be hundreds of meters (if not kilometers) thick and so unconsolidated that the dust would instantly swallow up any craft landing there. Geologists looking at the Moon suspected instead that the dust layer was at most only a few meters thick. They speculated that over geological time, the rock layers postulated to make up the crust had been ground up into a powder-like dust layer by the incessant micrometeorite bombardment.

Before taking the risk of landing men on the Moon, we had to understand the nature and extent of the lunar dust layer. Thus one of the prime scientific objectives of the robotic missions flown before Apollo was to understand the surface layer—how it formed and evolved over time—and to assess its potential risk to human missions. Giving this issue priority ensured that much effort would be expended trying to understand the distribution and origin of the lunar surface material.

Grind It Up and Glue It Together

The original crust of the Moon is made up of slowly cooled crustal rocks, volcanic lava, and blankets of impact ejecta. Most of these rock units formed over 3 billion years ago, very early in the Moon's history. Since then, the rocks have been exposed to space and, more important, to the constant bombardment of debris, ranging in size from specklike motes of dust to miniature planets. The population of objects hitting the Moon is dominated by the very small. The Moon is struck constantly by the very tiny, less often by the moderately small, and very rarely by the big. The meteoroid flux is dominated by dustlike particles; objects about the size of a car hit the Moon about once every 100 years. Dust-sized objects also strike Earth, but the atmosphere protects Earth's surface, and these particles burn up because of air friction *(meteors)* long before they get near the ground.

One might think that because most of the particles are small, they cannot do much damage. But these dust motes are very speedy, and most hit the Moon at velocities greater than 15 km/sec (at such a speed, North America could be crossed in about 5 minutes). This property of great speed means that impacting dust grains hit the Moon with a lot of energy, and it is this energy that makes craters. Think of the micrometeorite flux as a giant sandblaster, slowly grinding the Moon's crust into dust. This sandblaster has a factory defect—-occasionally it spits out a huge chunk of the Moon, digging out a rather sizable hole.

The ground-up layer of dust, rock, and debris ejected from craters completely covers the Moon (Fig. 4.1). Scientists call this debris layer the *regolith,* which a dictionary will tell you is "unconsolidated debris overlying bedrock." The regolith consists of *all* of the debris making up the surface of the Moon, not merely the finest dust. Thus boulder-sized rocks on a crater rim, hand samples picked up by the astronauts, and tiny specks that are too small to be seen and that turned the astronauts' spacesuits from pristine white to dull gray are all fragments of the regolith.

We first saw the regolith close up from the probes sent to scout the Moon for Apollo. From the hard-landing Ranger and

Plate 1. The Moon in "natural" color. Although often thought to be gray and colorless, it is basically a grayish-brown or red color. Some lava flows in the maria are decidedly bluer than the typical lunar color, a reflection of titanium in the rocks. Earth-based telescope photograph by and courtesy of Bill and Sally Fletcher.

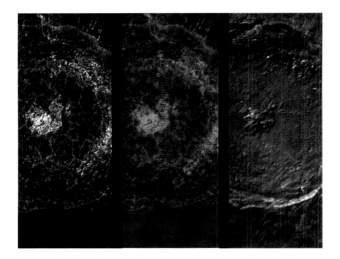

Plate 2. Three mosaics of the same scene over the crater Tycho (43° S, 11° W; 85 km diameter). At right is a "stretched" color mosaic of the crater, exaggerating the blue reflectance of the central peak. The false, multicolor version (center) further exaggerates these color contrasts and allows different rock units to be distinguished. The simple ratio image (left) shows the location of fresh, mafic material, that is. material relatively rich in iron (Fe) and magnesium (Mg). Such color data from the *Clementine* spacecraft will permit scientists to map rock types over the entire surface of the Moon.

Plate 3. The Imbrium basin on the Moon, portrayed as it would have looked immediately after its formation by the impact of a large asteroid 3.84 billion years ago. The glowing interior of the basin is made up of liquid rock, shock melted by the impact. Light "rays" of ejecta thrown out of the basin cover nearly the entire near side. Small volcanic eruptions show that the Moon was volcanically active at this time. Artwork by Don Davis.

Plate 4. Liftoff of the *Apollo 15* mission in July 1971. The massive *Saturn 5* booster rocket developed 7.5 million pounds of thrust on takeoff and lofted 120 tons into low Earth orbit.

Plate 5. Astronaut Jack Schmitt, the only professional geologist to go to the Moon, examining the rim of Shorty crater at the *Apollo 17* Taurus-Littrow site. Note, in the foreground, an exposure of orange soil; this material is volcanic ash that erupted from a lunar fire fountain over 3.7 billion years ago.

Plate 6. Microscopic view of the *Apollo 17* orange soil. It is made up of very small (each about 0.04 mm across) glass spheres; the orange color is caused by their high titanium content. These glass beads were formed during a fire-fountain eruption over 3.6 billion years ago.

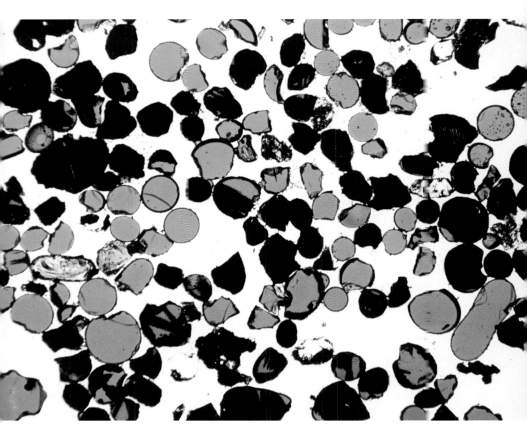

Plate 7. Sample of a mare basalt from the Taurus-Littrow Valley, the *Apollo 17* landing site. The thin-section image is a slice of rock, cut 0.03 mm thick and viewed through the polarizing filters of a microscope. Geologists do this to identify minerals, which have characteristic colors and shapes, and to show the texture of the rock, which gives us clues about its origin. This basalt is made up of the minerals plagioclase, pyroxene, and olivine.

Plate 8. False-color mosaic made with three bands of the ultraviolet-visible camera on *Clementine* of the Aristarchus Plateau. In this image, blues are fresh highland materials, deep red is dark pyroclastic (ash) deposits, yellow is outcrop of fresh basalt (in crater and rille walls), and reddish-purples are mare lava flows. The sinuous rille is Schröter's Valley, a large lava channel on the plateau. The mosaic shows the quality of the full-resolution data set from the Clementine mission.

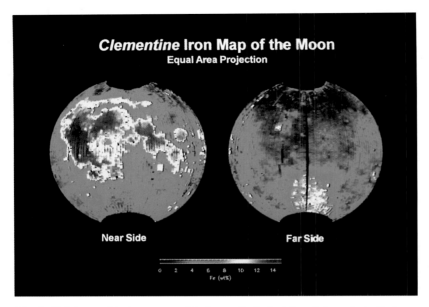

Clementine Iron Map of the Moon
Equal Area Projection

Near Side

Far Side

0   2   4   6   8   10   12   14
Fe (wt%)

Plate 9. Global map of the iron content of the lunar soil, obtained through the analysis of color data collected by the Clementine mission. The maria are made up of iron-rich basaltic lavas and so display soil compositions rich in iron. The highlands are very low in iron, supporting the concept that they are made up largely of the iron-poor, aluminum-rich rock type anorthosite. On the far side a large circular zone of relatively elevated iron content in the southern hemisphere is associated with the South Pole–Aitken basin, the largest known crater on the Moon. This zone is a result of the upper, iron-poor crust having been stripped off by this impact. Other areas of high iron content in the highlands are associated with ancient lava flows that have been buried by highland debris thrown out by the large basins.

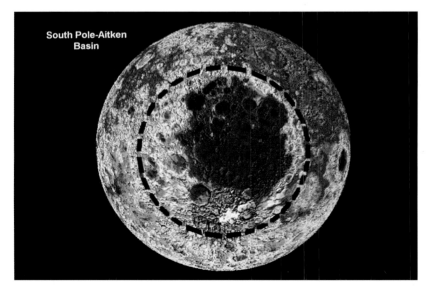

South Pole-Aitken
Basin

Plate 10. The South Pole–Aitken basin, over 2,500 km in diameter and the largest, deepest impact crater known in the solar system. This map shows altimetry data from the Clementine mission; colors represent elevation, with red being high and purple being low. The basin is over 12 km deep in some places.

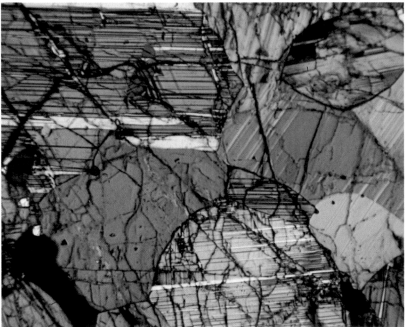

Plate 11. The rock type anorthosite, a slowly cooled, igneous rock made up almost entirely of plagioclase feldspar. Bottom: the thin-section view shows mostly plagioclase (gray) with very minor amounts of calcium-rich pyroxene (yellow). The composition of this rock and its abundance in the highlands indicate that the early Moon was nearly completely molten (the magma ocean).

Plate 12. Troctolite, a rock of the Mg-suite of the lunar highlands. This rock is made up of equal parts plagioclase (white) and olivine (yellow). Bottom: The rocks of the Mg-suite, formed from many different magma bodies, intruded into the crust over a very long time. Mg-suite rocks are common in the Apollo collections but not in the highland surface; they may make up large parts of the lower crust.

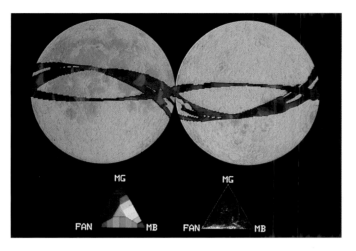

Plate 13. A rock-type map of the Moon made by analyzing data from the chemical-mapping instruments carried aboard the orbiting *Apollo 15* and *16* spacecrafts. In this method, blue indicates anorthosite, red indicates KREEP and the Mg-suite rocks, and green is the mare basalts. Results from this mapping, done before the Clementine mission, are consistent with the color data from that later mission, suggesting that the crust is almost entirely anorthosite, requiring an early magma ocean.

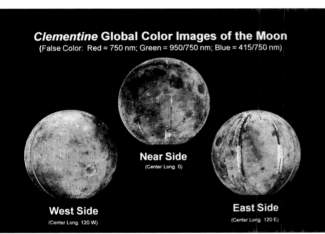

Plate 14. Global color data from the Clementine mission. The color data are shown at a resolution of one frame per pixel, or more than 300 times *lower* resolution than the actual data (to simplify and speed up image processing). The Moon displays many different color units, indicating a complex, heterogeneous surface composition. The predominance of green in the highlands indicates that much of the crust is made up of anorthosite, confirming earlier suppositions.

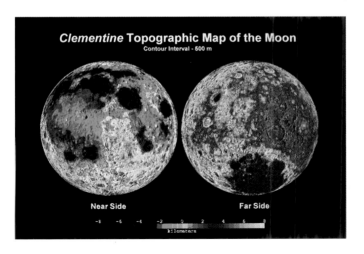

Plate 15. Global topographic map obtained from laser altimetry data of the Clementine mission (red = high elevations, purple = low elevations), showing the near and far sides of the Moon. Note the marked difference between the two hemispheres; the large (2,500 km diameter) South Pole–Aitken basin can be seen on the far side.

Figure 4.1. *Apollo 15* Lunar Module pilot Jim Irwin digging a trench in the regolith of the Hadley-Apennine Valley. Most of the regolith is very fine soil, although the term *regolith* includes the large rocks and blocks that are part of the debris blanket that covers the bedrock on the Moon.

the orbiting Lunar Orbiter missions, we observed that small craters had penetrated the upper layer of the Moon, throwing out blocks of the bedrock. Measuring the diameter of these craters allowed us to estimate the thickness of the surface layer. Because the regolith is formed by exposure to bombardment from space, the longer the time of exposure, the more regolith is made. Thus the thickness of the regolith increases with time. Very old areas have a thick regolith. In the highlands, the oldest regions of the lunar surface, the regolith may be 20 to 30 m thick. Mare sites, being younger, have thinner regolith, ranging

from about 2 to 8 m. The youngest rock surfaces of the Moon, the impact melt sheets that make up the floors of large rayed craters such as Tycho (Fig. 2.4), have the very thinnest regolith layers, in some cases only a few centimeters thick (Fig. 3.1).

Other parts of the Moon offer a different window to the regolith. The walls of large craters and rilles sometimes expose a complete section of the surface regolith *and* underlying bedrock (Fig. 4.2). During the *Apollo 15* mission in 1971 the astronauts were able to sample boulders that, if not in place as bedrock, are very close to the bedrock layer. All of the other Apollo samples come from the regolith, and our understanding of the regional geology of the Moon and the nature of the rock units that compose it is only as good as our ability to comprehend how the regolith forms and evolves. Fortunately, the Apollo missions returned samples that allowed us to study the regolith in detail, and this work provides us with most of our understanding of the complex nature of the regolith and the bombardment of the Moon.

The fragments that make up the regolith are nearly all derived from the rocks that underlie it. This mixture includes broken-up fragments of rocks, single grains of different minerals, small chunks of metal broken out of the rocks, and fragments of glass (Fig. 4.3). The mixing of materials by constant cratering, excavating, reexcavating, and burial produces a material that is extremely complex in detail but rather simple in bulk properties. For the mare areas, regolith is found to contain a considerable fraction of highland material, anywhere from 10 to almost 60 percent. In the highlands, mare material is found but at very low abundance, usually much less than 1 to 3 or 4 percent. This fact suggests that although the ejecta blankets of craters and their long rays may spread rock fragments far and wide, they are not efficient mixers of the two terrain types.

So where does all of the highland debris come from at the

Figure 4.2 (opposite): Exposed bedrock. Bedrock is exposed on the Moon where craters have removed the overlying regolith or where slopes have allowed gravity to remove the debris. In the walls of Hadley Rille, near the *Apollo 15* landing site (Fig. 3.6), the bedrock of the mare basalts is exposed, showing the layered nature of the maria near the site.

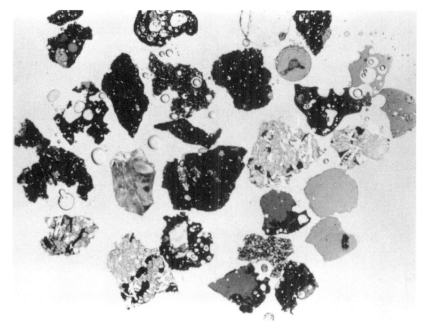

mare sites? All of the Apollo mare sites lie on top of fairly thin piles of lava flows. Thus it is easy for a crater to punch through the mare lavas and excavate the underlying highland terrain. We estimate that nearly all of the highland debris at mare sites is mixed into the regolith in this manner, that is, by the vertical mixing of underlying rocks (Fig. 4.4). Of course, lateral mixing by ray deposits surely occurs and is likely to be responsible for the chunks of mare lava found at, for example, the *Apollo 16* site (Fig. 3.8). The rarity of such fragments indicates the relative importance of this process of lateral mixing of crustal materials, a process that tends to be responsible for very much less than a couple percent of observed exotic debris. Put another way more than 95 percent of all material in the regolith is locally derived, usually from distances of less than a few kilometers away. One must always be aware, however, that a sample from any site on the Moon *could* come from nearly anywhere else on the Moon.

A Planet of Glass: Impact Melting and Agglutinate Formation

One would think that the net effect of regolith formation is entirely destructive: As the surface rocks are ground into a fine powder, the rocks become smaller and the dust becomes finer. In fact the regolith soon reaches a state where fine and coarse materials are in rough balance. This balance results from the fact that an impact can also *make* rock. When a projectile hits the Moon at cosmic speeds (an impact greater than about 5–10 km/sec is called *hypervelocity*), the shock wave generated by the impact will first compress and then rapidly decompress the target rock. This decompression results in material being vaporized and melted, leaving a zone of shock melt (or impact melt) that lines the crater cavity. For the very small craters, those that make up the vast majority of craters in the regolith, the film of melted rock lines the crater floor and may splash out into the surrounding terrain (Fig. 4.5).

Figure 4.3 (opposite). Top: a close-up view of the regolith, showing the fragments of rock and glass that make up the soil. Bottom: view of a thin section of soil fragments, cut 0.03 mm thick. The surface rocks have been slowly ground to this powder by micrometeorite bombardment over geological time.

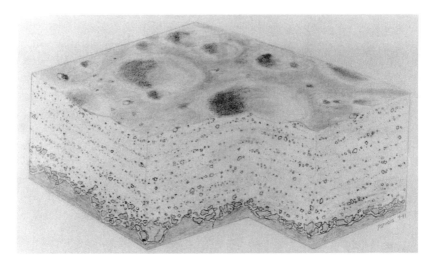

Figure 4.4. Block diagram showing the regolith overlying the shattered and fractured bedrock. The tendency is to make the grain sizes finer by impact grinding until a steady-state condition is reached.

These melt bodies are small, and they cool very quickly. A natural consequence of the quick cooling of molten rock is the formation of glass. Glass is material that does not possess internal structural order, unlike the crystals that make up minerals. The composition of the glass is determined by the composition of the melted target rock. An interesting property of impact melts is that they tend to homogenize diverse compositions of the crater target into a single melt composition. The glass formed during these small impacts can make up a significant fraction of the regolith. A rather surprising result of study of the *Apollo 11* sample is that over 60 percent of the soil returned from this first landing site is made up of glass.

Although small, clear spheres of impact glass are found at all sites on the Moon, the glass found in the regolith is seldom pure. It is often mixed with crushed rock and mineral debris from the local bedrock. We also find small specks of nearly pure iron in these glass fragments, making some of the glass magnetic. The intimate mixture of glass, mineral crystals, and tiny rock fragments into a chunk of glassy material is called an *agglutinate*

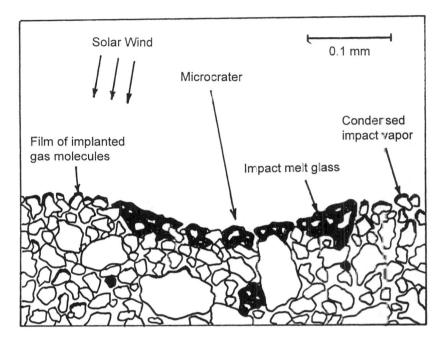

Figure 4.5. Close-up showing how agglutinate glass is created by the fusing (melting) of the mineral grains of the regolith. Gases from the solar wind are implanted onto the dust grains. After R. M. Housley, E. H. Cirlin, N. E. Paton, and I. B. Goldberg, "Solar Wind and Micrometeorite Alteration of the Lunar Regolith," *Proceedings of the Fifth Lunar Science Conference* (New York: Pergamon Press, 1975), 2623–42.

(Fig. 4.6). Agglutinates can compose a sizable fraction of the regolith, but their actual abundance varies widely. For example, agglutinates may be nearly absent on the rim of a relatively fresh impact crater that has excavated fresh bedrock from beneath the regolith. However, just a few meters away, agglutinates could make up more than 50 percent of the local material. Agglutinates are rare in the crater ejecta because these glass fragments form by small impacts and many are produced by the exposure of the regolith to space for long periods. Thus a large, fresh crater excavates material that has seen very little bombardment by micrometeorites.

Regolith becomes richer in agglutinates with time, a process called *maturation*. A mature soil has been exposed to space

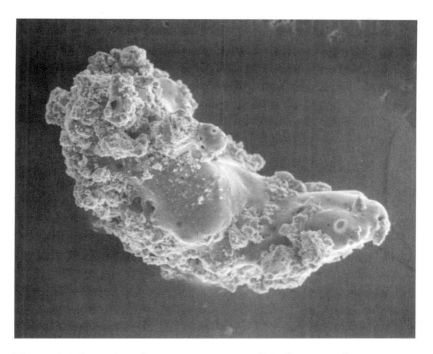

Figure 4.6. Scanning-electron microscope (SEM) image of an agglutinate from the regolith. Agglutinates are made up of glass, mineral, metal, and rock fragments, all bound together by impact-produced glass.

longer than has an immature soil and consequently contains many more agglutinates than does the immature material. The abundance of glass and agglutinate in a lunar soil is a measure of the length of time it has been exposed to space. These "maturity levels" are highly variable from place to place. In fact maturity can vary widely within the area of a single Apollo landing site (scales of a few kilometers). The concept of maturity is important in understanding remote-sensing data from the Moon because orbiting spacecraft sense the uppermost surface, which is almost exclusively regolith. In general the glass present in mature soils tends to mask the mineral absorption features in spectra that we depend on to map compositions from orbit (see Chapter 8). However, such information is only suppressed, not eliminated.

In detail, then, the regolith consists of a mixture of broken rock and mineral fragments, a variety of glasses (agglutinates and clear, homogeneous glasses of both impact and volcanic origins), blobs of pure metal (mostly iron), breccias of many types of rocks, and the occasional rare fragment of meteorite. It might be surprising at first that meteorite fragments are so rare. After all, the surface of the Moon is saturated by the holes left from meteorite impact. However, because the Moon has no atmosphere to slow down these fast-moving projectiles, they strike at cosmic velocities—nearly always at speeds great enough to vaporize most, if not all, of the impactor. This vapor is usually driven off the Moon (the lunar escape velocity is only 2.5 km/sec, a low value compared with Earth's 12 km/sec escape velocity), but occasionally the vapor recondenses onto the Moon. Some meteorites contain water, either bound in the crystal structure of minerals or as the ice component of the "dirty snowball" cometary fragment. This water is mostly lost from the Moon, boiled off during the heat of the lunar day. However, shadowed regions near the poles may be permanently dark, and if water molecules get into these dark, cold areas, they could remain on the Moon. Such deposition of water may have happened at the south pole (see Chapter 8).

Meteorite fragments in the regolith are preserved in two ways. Fragments occasionally hit the Moon at fairly low velocities; such objects may be nearly co-orbiting the Sun with the Earth-Moon system, thus encountering the Moon at relatively low velocity. An impact at such low speed could shatter the meteorite but not vaporize it. In the other method of preservation, fragments of a large projectile are split off at the exact moment of impact with the Moon. Some of these fragments might be cast away from the surface at speeds comparable to the impact velocity but in the opposite direction. Such a particle path would result in a net encounter speed (relative to the lunar surface) of a few kilometers per second or less. At such slow encounter speeds, fragments of meteorite would be preserved—not vaporized, as would be the main impacting body. These two possibilities would preserve meteorite fragments, but the fact that such fragments are exceedingly rare in the returned Apollo samples attests to the rarity of such events.

Sometimes we find specimens of rock and mineral fragments glued together by impact-melted glass. A rock made up of ground-up fragments of other rocks is called a *breccia*. The regolith from the *Apollo 11* site contained many of these samples; because the grain size was so fine, these rocks were given the special name *microbreccia*. In all regolith, we find a size continuum of impact-melted fragments, ranging from the microscopic glasses to the boulder-sized melt breccias. The regolith thus consists of both destructive material (the rock and mineral fragments ground up by the bombardment) and constructive material (the impact-melted and cemented agglutinates and breccias). So the formation of the regolith is a balance between impact destruction and reconstruction, between the crushing and breaking of an impact and the melting and fusing of the shock heat. In this bombardment dance, destruction greatly predominates, mainly because the total volume of material crushed and excavated during an impact greatly exceeds the volume of material that is melted and welded together. Thus, on balance, regolith formation is the disaggregation of a planet, and regolith thickness increases with time.

Regolith is found on all geological units on the Moon, and its formation has another important consequence: the erosion of surface features. In pre–space age movies and science fiction novels, the surface of the Moon was often depicted as an alien landscape of rough, craggy pinnacles with steep slopes, towering spires, and huge, yawing chasms (Fig. 1.7). The actual Moon is much smoother. The Apollo missions to the highlands displayed spectacular vistas of smooth, rolling hills extending to the horizon (Fig. 3.9). The bombardment by micrometeorites has smoothed the jagged surfaces, eroded the angular faces of rocks, and covered all but the very steepest slopes with fine powder. Such erosion can be seen in the rounding of some of the Apollo samples (Fig. 4.7). Although some writers claim that the real Moon is prosaic and duller than the rugged vistas of fiction, one beneficial consequence of impact erosion is that virtually all of the lunar surface is smooth enough to land on. We can explore easily on foot and by vehicle without the need for rappelling ropes and climbing pitons. Sometimes dullness can be a blessing!

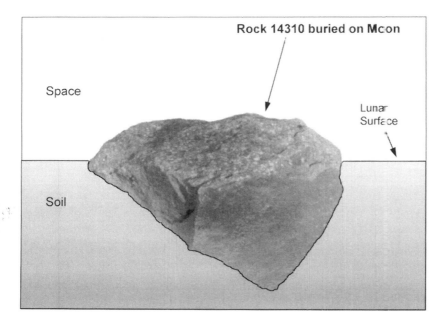

Figure 4.7. The two surfaces of rock 14310, showing the smoothed, rounded upper surface (eroded by micrometeorites) and its sharp, angular buried half (protected from meteorite erosion by burial). After Heiken, Vaniman, and French, *Lunar Sourcebook*, Fig. 4.17a.

Peppering the Surface through Time: History of the Cratering Flux

Because the rock units of the Moon are very ancient and it has no atmosphere, all sizes of meteorites have struck its surface throughout its history over 4 billion years. The Moon retains a record of this bombardment, a record that we can read if we are clever enough. The regolith can be used as a time probe to understand changes in the impact flux. As the assembly of the planets *(accretion)* ended, the composition of the projectiles that have hit the Moon (including meteorite types that no longer exist in the solar system) may have changed. The intensity of the impact flux as a function of time may have varied over the last few billion years, a topic with particular relevance to the history of life on Earth, as we shall see.

A major discovery of the Apollo missions was that the rate of

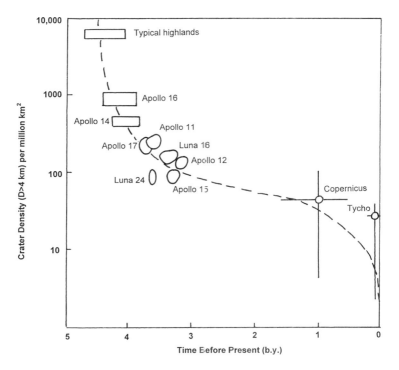

Figure 4.8. The impact crater frequency curve for the Moon. Note that the number of craters per unit area (N/km2) gradually decreases as surface ages get younger. The sampled points (e.g., Apollo 11, Luna 16) of the Apollo and Luna landing sites provide calibration marks that allow us to estimate the absolute ages of unvisited sites on the Moon. Simplified from Heiken, Vaniman, and French, *Lunar Sourcebook*, Fig. 4.15.

impact bombardment in the Earth-Moon system has not been constant over the history of the solar system. The early rate of bombardment was much greater than the current rate (Fig. 4.8). This decline in the impact flux is apparent through study of the regolith and is dramatically demonstrated by the comparison of rock ages from the highlands and mare. The most pronounced decline in the impact flux occurred between 4 and 3.8 billion years ago, about the time that the last major basins formed (see Fig. 2.20). However, between the time of the *Apollo 11* lavas (about 3.7 billion years ago) and the *Apollo 12* lavas (about 3.1 billion years ago), the flux was also declining at a greater rate

than it has since the *Apollo 12* lavas were extruded. General
rates of bombardment decreased by more than a factor of five in
this span of 600 million years.

Studies of impact-melted rocks show that a small fraction
(usually less than a few percent) of the melt composition represents a trace of the projectile that struck the Moon. Small traces
and the abundance of rare elements characterize certain meteorite types. Study of samples from the regolith can tell us if the
source of impacting objects has changed with time. For example, one suggestion is that the early Moon was sweeping up the
leftover debris from the accretion of the planets; thus this material would resemble the composition of the building blocks of
the planets. Later impacting objects came from farther out in
the solar system as cometary debris was perturbed into inner
solar system orbits by gravitational attraction of the giant outer
planets. Such a model implies that early impactors would have
had a composition different from that of the current crop of
debris. This idea can be tested by examining the types of projectile material in the regolith. Because early (that is, lower) layers
of the regolith contain material deposited from the former population, these data could be compared with projectile debris from
the younger (that is, upper) layers to determine if and how the
projectile compositions have changed with time. Initial efforts
to do this in a rather crude manner suggest that the populations
of impacting debris *are* different, but how these populations
might change in composition with time remains unresolved.

Melt products formed during the tail end of the heavy bombardment are presumably contained within the regolith layers
of the older mare sites. Although such samples have not been
recognized to date, they are extremely important to find. They
could address the question of whether the ancient impact flux
was made up of fragments different from those currently hitting the Moon. With only a few sites visited, we cannot be sure
of the rate of decline of the early impact flux, an important
issue in the general history of the Moon. The best evidence now
available suggests that the Moon underwent intense bombardment between 4.5 and 3.8 billion years ago, but the concept
that the early Moon experienced a "quiet" period, with a subsequent "cataclysm" forming all the observed large craters, cannot be excluded. (This topic will be discussed in Chapter 6.)

One of the most intriguing ideas in modern evolutionary research would, at first glance, seem to bear only the remotest relation to lunar science. The idea that a giant impact on the Yucatan peninsula of Mexico (the Chicxulub crater) 65 million years ago was responsible for the extinction of much of the life on Earth, including the dinosaurs, is now fairly well established. Scientists who study the fossil record and, in particular, the patterns of extinction through time have noted that this "Cretaceous-Tertiary" mass extinction is actually part of a larger pattern of crises in the history of life. Evidence from the fossil record suggests that major extinctions occur roughly once every 26 million years. One explanation is that showers of cometary and asteroidal debris pelt Earth at regular intervals. Even though Earth has witnessed the formation of many impact craters over its history, the dynamic processing of Earth's surface has erased many of them, leaving only a small number of craters exposed on Earth. Because of this active process of crater removal on Earth, the periodic impact idea is difficult to test.

This sounds like a job for—the Moon! The Moon has undergone the same bombardment history as our Earth. One glance shows an abundance of craters, literally everywhere. So how can this idea be tested on the Moon? If these mass extinctions are really periodic, occurring once every 26 million years, and if they are related to the impact flux, then crater ages on the Moon should cluster around certain times rather than show a smooth distribution. Mare sites on the Moon have been exposed to the impact flux for at least the past 3 billion years, roughly the length of time that life has existed on Earth. In impact craters, samples of the melt sheet can be dated by measuring the isotopic composition of the rocks. The radioactive isotopes that occur naturally in these rocks decay at a known rate, and the age of formation of melt samples can be calculated by measuring the concentration of these isotopes. By collecting samples of impact melt and dating them with this technique, we can reconstruct the impact history of the Moon and Earth.

The Apollo sample collection is quite limited, and we cannot determine, with the samples in hand, whether the impact flux varies periodically or not. However, if we return to the Moon, we can devise an easy experiment to seek out craters on the mare, collect samples of impact melt from their floors, date the samples, and see if they cluster at the proposed times in the geologic past. A finding from the Moon that such clustering of crater ages does not occur would pose grave problems for the model of an impact-related periodic extinction of species. By exploring the Moon, we can address a question of fundamental importance to the history of life on Earth.

Starprobe: Mission to Planet Dirt

Because the Moon has no atmosphere or global magnetic field, the stream of energetic particles, atoms, ions, and nuclei constantly spewed out of our Sun (the closest star) pours directly onto the lunar surface. Some of this sticks. From study of the first lunar samples, it was immediately known that light gases such as hydrogen and helium occur in lunar soil, although in exceedingly small amounts (for example, hydrogen is present at concentration levels of about 50 parts per million or less). The rocks contain very little volatile material, and it was soon realized that gases found in the regolith come from the solar wind. This outpouring of particles continually rains on the Moon.

Thus the gases contained within the regolith are mainly derived from the Sun. Hydrogen is the most abundant element, followed by helium, nitrogen, and the inert gases neon, argon, and krypton. Other light elements of at least partly solar origin are also present, including carbon and sulfur. The regolith has formed over a 3-billion-year period, and it contains a record of the output of the Sun over geological time. In the same manner that the ancient impact flux may be studied and characterized through the study of the regolith, we can also reconstruct the ancient Sun and the gases and particles it has spewed out over its lifetime. Studies of the Apollo regolith samples show that the ancient Sun (before about 3 billion years ago) had a solar wind whose composition was slightly different from that of today's Sun. In particular, the ratio of the two principal types of nitrogen (^{15}N to ^{14}N isotopes) was higher in the geologic past than it is today. Such a relation is neither predicted nor explained by current models of stellar evolution. This rather startling result suggests that the processes of nuclear fusion in the core of the Sun are not fully understood.

The presence of hydrogen implanted by the solar wind on lunar dust grains is a finding of enormous consequence for the future exploration and use of the Moon. Although of very low concentration, there is enough hydrogen in some regolith to extract and use. For example, a square patch of mare soil 1 km on a side and 1 m thick contains enough implanted hydrogen to power a launch of the Space Shuttle, which uses a mixture of liquid hydrogen and liquid oxygen for fuel. Although lunar hydrogen is seemingly of low concentration, industrial-scale processing on Earth often mines material of far lower grade.

The hydrogen contained in lunar soil can be extracted by heating the soil to a temperature of about 700°C. Such thermal environments can be produced by the simple focusing of sunlight by a curved mirror (solar thermal furnaces). In addition other implanted volatile elements, such as nitrogen and sulfur, are also liberated by the thermal processing. The movement and the handling of such enormous quantities of material would pose considerable engineering problems, but they violate no basic principles of physics. We will examine the nature and potential uses of lunar resources in another chapter (see Chapter 9). In brief the production of water and rocket fuel on the Moon (the two principal uses of lunar hydrogen) would completely change the economics and dynamics of operating a settlement on the Moon.

Another gas derived from the solar wind may also be useful in the future. Some scientists think that a rare isotope of helium (^{3}He, whose nucleus contains only one neutron instead of the two neutrons carried by typical helium atoms) may be used as fuel in advanced fusion reactors to produce commercial power here on Earth. Although ^{3}He is extremely rare on Earth, it occurs naturally on the Moon at about the solar abundance, relative to hydrogen (at the still tiny concentration of about 4–5 parts per *billion* of ^{3}He in mare soils). The mining of ^{3}He would be similar in concept to hydrogen mining, that is, solar thermal heating of the regolith to liberate the adsorbed solar gases. The scale of operations, however, would be much larger, owing to the extremely small concentration of ^{3}He. For example, at 100 percent extraction efficiency, 200 million tons of soil would have to be processed to get 1 ton of ^{3}He. Such a mining prospect corresponds to an area 10 km (about 6.25 miles) square, strip-mined to a depth of 2 m (6 feet). To extract enough ^{3}He to provide the energy needs of the United States for a year, 25 tons of ^{3}He would be needed. Thus we would need to strip-mine an area on the Moon roughly 50 km square (to a depth of 2 m), which roughly corresponds to an area about the size of metropolitan Chicago. Such a mining prospect would be visible from Earth through a telescope but would not significantly alter the appearance of the Moon. The fusion reactors in which the ^{3}He would be used do not yet exist, even conceptually. However, within the next 50 years or so, the commercial production of power using ^{3}He fusion is well within the realm of possibilities.

The regolith also provides us with a window into the history of the galaxy. Very high energy particles (cosmic rays) directly interact with the lunar dust. Remnants of exotic isotopes produced during the deposition of material thrown out of *supernovae* (stellar explosions in the galaxy) may be found within the regolith. The cosmic ray bombardment creates short-lived radioactive isotopes, which allow us to determine how long a given rock or soil sample has been exposed to space. Such information makes it possible for us to reconstruct the history of a given rock sample to an incredible degree of detail. Yet another aspect of the continual cosmic ray bombardment is the interaction of cosmic rays with certain elements. This bombardment results in the reemission of gamma-rays that have energy characteristics unique to the element. Such gamma radiation can be detected from orbit, allowing us to map the distribution of several elements of geological interest with a gamma-ray spectrometer. The cosmic ray irradiation of the Moon actually aids us in our relentless probe of lunar secrets.

The ground-up surface layer of the Moon holds a wealth of information about the Sun, the solar system, the galaxy, the evolution of life on Earth, and of course the Moon itself. These aspects of the regolith make it an incredibly rich subject for investigation and a prime objective for study when we return to the Moon. Work on the existing Apollo samples continues to yield new insights into a variety of problems. All of this, plus the promise of future resources we can use to help us explore the Moon and near-Earth space, arises from a dusty pile of rubble in space.

 Chapter 5

The Fire Inside
Volcanism and Tectonism in the Maria

To the casual observer, the dark maria of the Moon are very striking. These smooth, low, dark plains occupy about 16 percent of the lunar surface area, but because most are on the near side, they appear to make up a much higher fraction. A variety of ideas about the origin of the maria flourished in the days before the space age. As models for the maria, such concepts as dried-up riverbeds, huge bowls of dust, flows of volcanic ash, or melted material ejected from basins, the largest craters on the Moon, were all proposed. However, in his landmark 1949 book, *The Face of the Moon,* Ralph Baldwin presented convincing evidence that the maria were floods of basalt, a common, dark, iron-rich lava that is abundant on Earth.

Baldwin's supposition did not remain unchallenged, and it took the detailed examination of the Moon in preparation for the Apollo missions, as well as the return of samples of the maria, to settle the issue. The maria indeed are made up of floods of lava, but what is most striking from the returned samples is the age of these lavas. The basalts returned by Apollo range in age from 4.3 to 3.1 billion years old, as old as the very oldest rocks on Earth. (Earth itself is 4.5 billion years old, although rocks of that age have not been found.) The ages of the mare basalt attest to the extreme antiquity of the Moon's surface. The returned lavas also show some interesting chemical characteristics, including a complete absence of any type of mineral that contains water, minerals that nearly always occur in lavas on Earth. These properties, determined during the initial examination of the Apollo samples, gave us a first-order understanding of the basic properties of the Moon.

Lunar Lava Flows

From pictures taken by telescopes on Earth, the maria appear smooth and dark. The impression is that these plains fill in the holes and depressions of the Moon, suggesting a fluid emplacement. At close-up scales small, lobe-shaped scarps can be seen; such scarps are very common in the lava flows of basalt found on Earth (Fig. 5.1). These scarps can even be seen in some of the best telescopic pictures of Mare Imbrium. Thus *before* the exploration of the Moon by spacecraft, we had direct evidence for emplacement of the maria by fluid flows. Needless to say, such evidence did not convince the unbelievers; indeed, a few of them remain unconvinced.

We obtained our first detailed look at the maria from the robotic precursor missions sent to scout the Moon for Apollo. The *Ranger 7* spacecraft was the first to return close-up pictures of the Moon. From these images, we learned that the scale of impact cratering continues downward to the limits of resolution and that the maria are covered by the regolith. Lunar Orbiter spacecraft took detailed pictures showing us a wide variety of landforms most easily attributable to volcanism, including a better view of flow fronts, small domes and cones, snakelike rilles that served as conduits for molten lava (lava channels and tubes), and irregular craters whose shapes are difficult to explain by impact origins. *Surveyors 1, 3, 5,* and *6* all performed soft landings in the maria, giving us a close-up view of the surface and telling us about the nature of the mare surface, including the revelation of some dark rocks covered with small holes, a morphology typical of lava samples.

The samples of the maria returned by the Apollo missions are a form of common lava known as *basalt*. Basalt is a dark lava, made up partly of minerals rich in iron (thus accounting for their dark color) and magnesium. The grain size of basalt is very fine (usually less than 1 mm), a result of the very rapid cooling (Plate 7). Like some Earth basalts, some lunar lavas have small, bubble-like holes *(vesicles)*, indicating that the magmas contained gas during eruption. The first basalts returned from the Moon were from the *Apollo 11* landing site in Mare Tranquillitatis. These rocks are remarkable in several respects. The

Figure 5.1. Lava flows on Earth and the Moon. Top: a flow of basalt lava in Hawaii, showing a leveed channel within it. Note the lobate margins of the flow. Bottom: flow lobes within Mare Imbrium on the Moon. The flow at top (arrow) displays a leveed channel, similar to the Hawaiian lava flow.

lunar lavas not only are devoid of water or hydrous phase (remember that the hydrogen found in the soil is from the solar wind, see Chapter 4) but also are depleted in all of the volatile elements (those that have very low boiling temperatures), including sodium, zinc, potassium, and phosphorous. Strangely (and surprisingly), the *Apollo 11* basalts have large amounts of titanium, mostly in the form of the mineral ilmenite, an oxide mineral of iron and titanium. The enrichment of the mare basalts in iron and their depletion in aluminum, the exact reverse of the composition of rocks from the highlands (see Chapter 6), account for the relative darkness (low albedo) of the maria as opposed to the terrae.

Lavas from the Moon were found to contain some minor minerals that are not found in Earth rocks. One of these, another iron-titanium mineral, was given the name armalcolite, in honor of the *Apollo 11* crew (the word comes from the first letters in the names of the crew: *Arm*strong, *Al*drin, and *Col*lins). The compositional properties of the lunar basalts reflect the unique chemical environment in which they formed: a small planet (resulting in low interior pressures) depleted in volatile elements, containing no water, and erupted onto a low-gravity surface in a vacuum. The mare basalts are extremely old by terrestrial standards. Basalts from *Apollo 11* crystallized as a series of lava flows that erupted between 3.8 and 3.65 billion years ago. For comparison, the largest areas of Earth covered by basalt are the floors of the ocean basins. These basalts range in age from zero to about 70 million years old. So the mare lavas on the Moon, some of the youngest lunar rocks, are at least 50–100 times older than comparable rocks from Earth! Basalts *were* being erupted on Earth before 3 billion years ago, but such rocks have been destroyed by terrestrial geological activity.

The chemical composition of the mare basalts has an interesting side effect. In systems of melted rock (magma), the *viscosity* (how "runny" a liquid is) of the liquid depends on the composition and temperature of the magma. The low amounts of aluminum and alkali elements and the high amount of iron in the lunar magmas, coupled with their relatively high temperature at extrusion, result in lavas that have extremely low viscosity. The viscosity of erupted lunar lava is about the same as motor oil at room temperature, much more fluid than terrestrial lava. Such runny, fluid flows spread out great distances, and this property, in addition to the low lunar gravity, accounts for the great lengths (up to hundreds of kilometers) that lava flows can reach on the Moon. Such a fluid character to the lava also explains the tendency of mare lavas to form low, broad structures rather than steep-sided volcanoes and to be erupted in lava channels, as are many basalt lava flows on Earth, such as the volcanoes of Hawaii.

Mare basalts from the other Apollo missions largely confirm the initial impressions gathered from the study of the first rocks brought back from the Moon, with some surprises and interesting variations. Lavas from the second mission, *Apollo 12*, are lower in titanium than the *Apollo 11* basalts and 600 to 700 million years younger (*Apollo 12* lavas formed about 3.1 billion years ago). Once again, these lavas are low in volatile elements and very rich in iron. The lower titanium and younger ages of the *Apollo 12* basalts confirmed that the maria were not erupted as a single, massive flood of lava across the surface all at one time but rather were formed in an extended process that involved different batches of magma erupting in different places at different times. In short, the samples told us about the complicated volcanic history of a small planet with its own geological evolution.

Mare basalts from the other two mare landing sites extended our picture of mare volcanism, perhaps somewhat misleadingly. *Apollo 15*, which landed just inside the rim of the basin containing Mare Imbrium (Fig. 3.6), returned low-titanium basalts of slightly older age than those from *Apollo 12;* these lavas crystallized about 3.3 billion years ago. *Apollo 17*, landing on the edge of Mare Serenitatis, returned very high titanium basalts, similar to those from *Apollo 11* but of a different age, about 3.7 billion years old. These results suggested to some scientists that the Moon had a fairly simple volcanic history, with early eruptions of high-titanium lavas and late eruptions of low-titanium lavas. The conclusion was also drawn that the Moon "died" volcanically after the *Apollo 12* lavas erupted at 3.1 billion years, a totally unwarranted conclusion that even today persists in lunar folklore.

In addition to the basalt samples returned from the mare

landing sites, little fragments of lava have been found in samples from the highlands as well. These basalts occur in two principal ways: as small rocks in the regolith of highland sites and as fragments of volcanic lava in highland breccias. The former occurrence of mare basalt is likely to result from the deposition of a ray from a crater on the maria, distant from the highland site. Examples are several fragments of high-titanium basalt from the *Apollo 16* regolith. Because secondary craters from the large impact crater Theophilus occur near the site (Fig. 3.8), we infer that these rocks were thrown to the site by the formation of this crater and that they represent a sample of the distant Mare Nectaris. So we can characterize the rocks of distant maria if plausible candidate craters can be identified.

In contrast, mare basalt fragments in highland breccias offer clues to the variety and ages of the earliest phase of lunar volcanism. These breccias from the highlands were assembled before 3.8 billion years ago; therefore, the lava fragments within them must be older than this. Some of the mare basalts are large enough to date by measuring their radioactive isotopes. We find that mare lavas were extruded well before 3.9 billion years ago. The oldest mare basalt yet found is about 4.2 billion years old, only slightly younger than the age of the solidification of the crust. Other fragments date from between 4.1 and 3.9 billion years and display a variety of chemical compositions. Curiously, most of the ancient mare basalt fragments tend to have relatively high contents of aluminum compared with the basalts from the "main phase" of mare eruptions, although a few groups of high-aluminum basalt date from this later era as well.

Basalt is created by partially melting rocks composed mostly of the iron- and magnesium-bearing minerals olivine (a green mineral whose gem version is known as peridot) and pyroxene (a mineral group that includes jade on Earth). From the relatively high density of the mantle, we know that it is largely made up of the minerals olivine and pyroxene. Radioactive, heat-producing elements, such as uranium, made the early mantle very hot—in some places, hot enough to partially melt. These blobs of melt coagulate deep in a planet's interior and slowly migrate upward, where they may force their way to the surface and be extruded onto a planetary surface as a lava flow.

The chemistry of basaltic magmas tells us approximately where they formed within the Moon and what processes have affected them subsequently. Our study of the mare basalts reveals that many regions of the mantle underwent melting episodes at several depths over a very long period of time, a period at least 700 million years long and more likely 1–2 billion years long. These melted pockets found their way to the surface through cracks that they propagated or through the fractures induced by the formation of the giant craters and basins of the highlands. However, only a very tiny percentage of the mantle has been melted to make basalt. Although the maria appear prominent (Fig. 2.1), the lavas are relatively thin compared with the volume of the crust as a whole. It is estimated that the mare basalts probably account for less than 1 percent of the total volume of the crust.

Fire Fountains: Ash Deposits on the Moon

Both the *Apollo 15* and the *Apollo 17* missions returned some unexpected volcanic material. Small glass beads were found in abundance at both sites: clear emerald-green glass at the *Apollo 15* site and black-and-orange glass from the *Apollo 17* site (Plate 6). These glass samples are homogeneous in their basaltic composition and do not contain the debris of mineral fragments that characterizes the impact-melted agglutinates from the regolith. The surfaces of these glass beads, which have been studied by electron microscope, have small glassy mounds (Fig. 5.2) made up of a variety of volatile elements including lead, zinc, and halogens such as chlorine. The *Apollo 15* green glasses are very rich in magnesium and extremely low in titanium (an unusual composition for a lunar magma), whereas the orange-and-black *Apollo 17* glass is rich in titanium. Once these glasses were recognized from these two sites, where they occur in abundance, small varieties of similar material were recognized at every other landing site. More than twenty varieties of volcanic glass are currently known.

Thus these glasses are of volcanic, not impact, origin, and they represent the products of a spray of low-viscosity lava into space. In Hawaii, eruptions of lava are sometimes accompanied

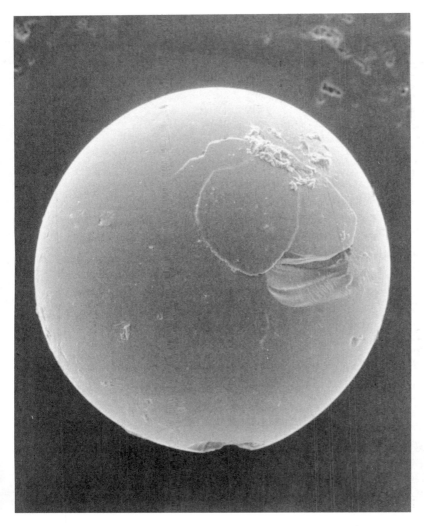

Figure 5.2. SEM image of the surface of a volcanic glass sphere, showing coating by mounds of material. These coatings are made of volatile elements such as sulfur, zinc, and lead and are additional evidence for the origin of these glasses in volcanic fire fountains.

Figure 5.3. A fire fountain of lava in Hawaii. Such eruptions are caused when a volatile-rich basaltic magma is released from a relatively small source vent under high-driving pressure, creating a spray of fine lava droplets. On the Moon such eruptions produced the dark mantle deposits seen in some areas (Fig. 2.16). U.S. Geological Survey photograph.

by very large sprays of magma from the vent. Such spray eruptions are called *fire fountains* (Fig. 5.3) and result in a deposit of ash around the eruptive vent. The ash from Hawaii consists of glass that has a basaltic composition and is frequently coated by a layer of volatile elements. On the basis of similar characteristics, we infer that the lunar glasses represent the products of fire fountains that existed on the Moon over 3 billion years ago. One difference between the lunar and the Hawaiian ash deposits is that so far no samples of lava corresponding to the lunar ashes in composition and derived from the same magma have been recognized.

The glasses are also evident as regional deposits. During the systematic mapping it was noted that parts of the highlands and maria are blanketed with a very dark material (Fig. 2.16). In the early days of lunar mapping, darkness was often equated with geological youth, and these regional dark deposits were thought to represent the ash deposits of young volcanism. In fact such a concept was responsible in part for the selection of the *Apollo 17* landing site near the margin of one of these regional deposits. It was predicted that this material was volcanic ash because it typically occurs in the vicinity of irregular craters of volcanic origin and indiscriminately covers all previous terrain. The principal occurrences of these regional deposits are around the margins of filled mare basins, such as those on the Aristarchus Plateau (Plate 8), Sulpicius Gallus, Rima Bode II (Fig. 2.16), and Taurus-Littrow (the *Apollo 17* landing site). The *Apollo 17* ash is indeed of volcanic origin, but it is old (3.5 billion years), not young.

Several large craters on the Moon have deformed and fractured floors. Along some of these fractures are small (typically a few kilometers in size) irregular craters surrounded by a dark, smooth material (Fig. 5.4). These craters are probably volcanic vents surrounded by ash deposits. They are the lunar equivalent of the *cinder cones* found in terrestrial volcanic fields. To create a cinder cone, magma from depth is squirted out through a very narrow conduit. The release of the low-viscosity lava under high pressure through a small vent causes the lava to spray into a "mist" of liquid-rock droplets. The spray of droplets quickly cools in flight, and each droplet lands back on the Moon as a small bead thrown on a ballistic path, like a pebble launched from a slingshot. Millions of such beads are made during an eruption, building up a deposit of dark ash that surrounds the vent.

Phase chemistry allows us to determine which minerals can coexist at certain temperatures and pressures. Studies of the phase chemistry of volcanic glasses have given us a great deal of insight into the very deep interior. It appears that these glasses

Figure 5.4. The crater Alphonsus (119 km diameter), a fractured-floor crater that has small, dark-halo craters of volcanic origin. These craters may indicate an intrusion of basalt beneath the crater floor.

were generated by the partial melting of an olivine-rich mantle at depths of about 400 km within the Moon. Moreover, unlike all of the mare basalts, the glasses appear to be largely unmodified from their chemical composition at their point of origin. Such a relation indicates that the magmas from which the glasses formed must have ascended very rapidly up through the Moon from deep within the mantle, with little chemical modification from their point of origin. The glasses then erupted into a violent spray at the surface. As such, lunar volcanic glass is our best sample of the deep interior and is an important material for determining the bulk composition of the Moon.

Understanding the mantle is one of the Holy Grails of lunar

science. Although the phase chemistry of the volcanic glass is the best sample of this remote region recognized to date, finding an actual chunk of the material that makes up the mantle would be the best way to comprehend this remote and inaccessible region. The rapid movement of magma through a large section of the Moon during the eruption of the glasses suggests another possibility. Small fragments of the surrounding rock may be ripped from the walls of the conduit through which the magmas rise. Such fragments (called *xenoliths,* meaning "stranger rock," because they are nearly always of a composition very different from the host rock) are found in some lava flows and ash deposits on Earth. It is possible that xenoliths from the mantle might someday be found in the ash deposits of the Moon.

Both the fountain nature of the eruption and the small coatings of volatile materials on the glass surfaces (Fig. 5 2) indicate that pockets of gas and other volatile elements existed deep with the Moon during the main era of mare volcanism over 3 billion years ago. Such an inference is also supported by the vesicles that are found in some samples of mare basalt (Plate 7). The composition of this gas phase is something of a mystery. It certainly is not water vapor (which accounts for almost all of the volatile phase in terrestrial volcanoes) because there is a total absence of water-bearing phases and oxidized material in the mare basalts. The reduced chemistry of lunar lavas makes us think that the gas phase might have been carbon monoxide. A detailed search for trapped bubbles of ancient gas within the glass spheres has not been successful to date, but the quest continues.

Basin Filling and Lava Flooding of the Moon through Time

As mentioned above, the maria were not erupted all at once as a massive flood of lava. The Moon underwent a long and pro-tracted volcanic evolution, characterized by different degrees of interior melting and different types of eruptions of different compositions at different places over a long period of time. The production of magma through time is an important basis for reconstructing lunar thermal evolution. Such information allows us to compare the Moon with the other terrestrial planets to understand the many ways that planets lose their heat.

Discrete, single flows of mare basalt appear to be rare. Although many units in the maria have a uniform density of impact craters (denoting continuity of age) and a single color, they appear to be made up of many thin, small lava flows. High-resolution photographs sometimes reveal scarps or moats occurring around impact craters; these scarps may delineate very thin flow lobes (less than a couple of meters thick). The spectacular lava flows within the Imbrium basin (Fig. 5.1) are often used in textbooks to illustrate the volcanic nature of the maria, but these flows are unique on the Moon, and their appearance probably indicates a specialized set of eruption circumstances (e.g., the rapid inflation of the crust and the discharge of a large magma body over a short period of time). For the most part, the maria appear to form a smooth, nondescript surface, and discrete flows are not visible. It is likely that the maria consist of a complex series of relatively thin lava flows, in which subsequent regolith production and impact erosion have completely destroyed any original volcanic texture.

Eruptions of mare basalt were rare events, even at the height of lunar volcanic activity, between 3.8 and 3.0 billion years ago. Although the maria appear to dominate the Moon, especially on the near side, the visible mare deposits make up much less than 1 percent of the volume of the crust. The total accumulated thickness of lava in most mare deposits varies widely but is typically less than a few kilometers, and large areas of basalt may be thinner than 100 m. In part we know this because of the abundance of highland debris mixed into the mare soils. As mentioned in Chapter 4, this fraction may approach 60 to 70 percent. Because most of this debris is derived from rocks beneath the local bedrock, the implication is that the stack of mare flows is thin.

There is a tendency to think of the maria as a hotbed of geological activity on the Moon, at least in the past. Certainly there has been activity in the maria, but consider the time spans involved in this activity. At the *Apollo 11* site, several different lava flows are represented among the samples. The oldest flows are 3.86 billion years old, but samples of flows of similar composition have a variety of ages, some as young as 3.55 billion years. In addition, lavas in another group, of different composition, are also about 3.5 billion years old. Thus, at this one site, we have evidence for at least four (and perhaps more) separate lava flows, emplaced over a period of more than 0.3 billion (300 million) years. This geologically "active" area of the Moon has been completely quiet for a period of time longer than vertebrate life has existed on Earth, except for a few months of activity. If this isn't the epitome of inactivity, what would be?

The very long lengths of time between the extrusions of individual lava flows suggest that these ancient flows may have been exposed to space for extended periods. Thus each basalt flow probably had enough time to develop a layer of regolith on its surface. Subsequent flows might have covered this ancient regolith with a layer of lava. Such a burial opens an exciting possibility: If buried regolith could be found, studies of the dust grains in them would give us a snapshot of the radiation and particle output of the Sun and the galaxy, not as it is now but as it was over 3 billion years ago. The regolith is a "time capsule" of extraordinary magnitude! Layers of ancient regolith could be accessed through exposure in crater, rille (Fig. 4.2), or graben walls—anyplace that layers of lava (the *stratigraphy*) might be exposed.

The volume of material erupted during a volcanic episode determines the shape of the resulting landform. Massive eruptions of large volumes of magma produce long, tongue-shaped lobate flows, not unlike the flow of molasses poured out onto a tabletop. Such eruptions typically are emitted from long, slot-like vents that permit great volumes of lava to reach the surface quickly. Smaller volumes of lava that come out from a narrow, more localized vent (such as a single-pit crater) can produce a variety of other interesting landforms. If the volume of lava erupted cools on timescales similar to the rate of magma supply to the vent, small volcanoes form and may assume a variety of different shapes (Fig. 2.15). Typically, lunar volcanoes form domes of low relief, a few hundred meters high and a few kilometers across. Such landforms resemble the small basaltic shields found in certain volcanic regions of Earth, such as Iceland and the Snake River plain of Idaho.

Other domes appear to be slightly larger and steeper (Fig. 5.5). One of the most spectacular areas in the maria is the Marius Hills,

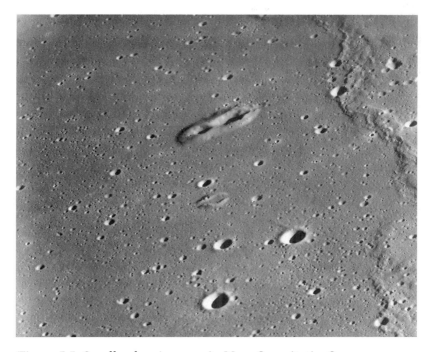

Figure 5.5. Small volcanic cones in Mare Serenitatis. Cones are typically made of cinders and ash, but on the Moon, cinder cones would have a very low profile because of the great distance material is thrown in the low lunar gravity.

the complex area of many small domes in Oceanus Procellarum mentioned in Chapter 2 (Fig. 5.6). The domes of the Marius Hills appear to be slightly steeper than the basalt shields mentioned above. On Earth such differences in shape are caused by differences in the composition of the lava, with steeper domes containing more silica and less iron and magnesium than do the low, broad shield volcanoes. The causes for steep domes on the Moon are less well understood but appear to be related to styles and rates of extrusion rather than to lava composition. Eruptions of shorter duration, possibly mixed with minor interludes of ash eruption, would build up a construct with steeper slopes than would the quiet effusion of the very fluid, low-viscosity lavas.

Domes and cones on the Moon are seldom found in isolation but often occur as fields of volcanoes within the maria. The

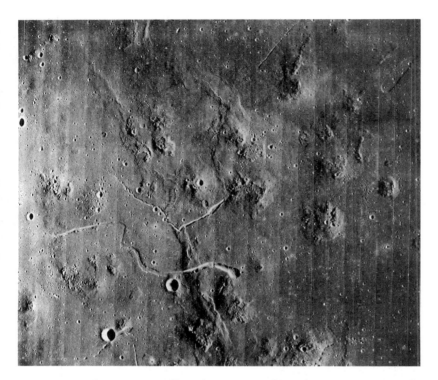

Figure 5.6. The Marius Hills volcanic complex. This area consists of a series of cones, domes, ridges, and sinuous rilles, all situated on a broad, regional upwarp in the maria. It may be the lunar equivalent of the large volcanic shields we see on other terrestrial planets, such as Olympus Mons on Mars.

Marius Hills display many domes and cones, which occur on the summit of a broad topographic swell. The complex is several hundred meters high and has a blister-like profile (Fig. 5.7), suggesting that it may be the lunar equivalent of shield volcanoes found on Earth, Venus, and Mars. The Rümker Hills in northern Oceanus Procellarum form another volcanic complex, similar in appearance to the Marius Hills but smaller. Small basaltic shields, such as those found near Hortensius (Fig. 2.15), occur in several locations near the margins of Maria Imbrium, Nubium, and Serenitatis. The last type of central-vent volcano on the Moon is the cinder cone, typified by irregular, dark-halo craters found along fractures in some crater floors (Fig. 5.4).

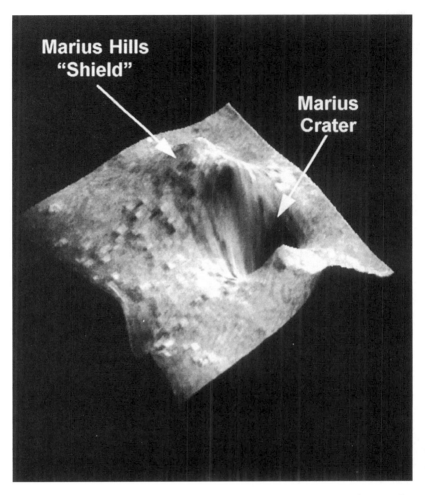

Figure 5.7. Three-dimensional, perspective topographic rendition of the Marius Hills based on Clementine laser altimetry. The shape of volcanic field suggests that this area is a lunar shield volcano.

These features are surrounded by ash deposits (as shown by remote-sensing data) and are often associated with the volcanic modification of large craters that originally formed by impact.

One common landform of the maria deserves special mention, if only because the *Apollo 15* mission (July 1971) was sent specifically to investigate one of them (Fig. 3.6). Sinuous rilles are narrow, winding valleys that occur primarily within the maria.

Some originate in highland terrain, but all trend downslope and empty into mare material. Many rilles begin in irregular craters (Fig. 3.6), some of which are surrounded by dark mantle material. Before the Apollo missions, many ideas were advanced for the origin of these features, including their origin as water-cut stream channels. However, the absence of any water on the Moon, the basaltic nature of the maria, and the irregular shapes of these features all have led to the consensus that rilles are lava channels, some of which were partly roofed over to form lava tubes (Fig. 5.8).

Figure 5.8. Hadley Rille, a lava channel-tube in Mare Imbrium, site of the *Apollo 15* landing. Evidence from Hadley Rille confirmed that sinuous rilles on the Moon are features created by flowing lava.

The *Apollo 15* mission was sent to Hadley Rille, just inside the rim of the Imbrium basin (Fig. 3.6). Hadley is one of the largest sinuous rilles on the Moon, being over 100 km in length, 1 to 3 km wide, and up to 1 km deep. The rille begins in an elongate, irregular crater in the Apennine Mountains, winds its way through the maria, snaking back and forth through the maria, and finally becomes shallow and appears to merge into a complex set of fractures north of the *Apollo 15* site (Fig. 3.6). The rille was examined at its rim near the landing site, where the rille is 1.5 km wide and 300 m deep (Fig. 5.8). Samples of basalt collected on the rille edge are probably the only samples in the Apollo collections that were taken from bedrock. A ledge of bedrock seen in the orbital photographs probably consists of this mare basalt unit (Fig. 4.2). Layers of mare lava are exposed in the walls of Hadley Rille, with all layers confined to the upper 60 m of the rille. Color data from the Clementine mission confirm that the walls of sinuous rilles on the Moon expose mare basalt (Plate 8).

Although all scientists agree that sinuous rilles are lava channels and tubes, the exact mode of formation remains somewhat contentious. On Earth, lava channels are created when a flow that is extruded at moderate rates cools from the margins inward; this cooling tends to confine the molten, active part of the flow along a central axis. This axis becomes the channel and in some cases is bridged over to form a lava tube. In this mechanism, lava channels are primarily *constructional* features, in which the overflow of lava builds up levees and raises the topographic level of the flow axis. Lava can accumulate laterally on the walls of the channel, narrowing the channel width; in fact such narrowing away from the vent is a common feature both of lava channels on Earth and of sinuous rilles on the Moon. Another concept that has emerged in the last few years holds that lava channels are primarily *erosional* features. The claim is that the eruption of very high temperature, fluid lava would flow turbulently and would soften, melt, and then remove underlying material, forming a lava channel by erosion. In such a model the sinuous depression in the maria consists of material removed by the flow of liquid lava and incorporated into the mare deposits.

The occurrence in the highlands of rille source craters that are

connected to sinuous rilles by channel segments in terra material indicates that some erosion has occurred. This eroded seg-
ment may have been enlarged by collapse, a process common in
lava channels. In general, however, geological evidence indicates that sinuous rilles and terrestrial lava tubes and channels
are formed dominantly by construction. The large size of sinuous rilles, argued by some as evidence for erosion, can also be a
result of the filling in of preexisting depressions, as clearly
shown by Hadley Rille, where the lava channel merges into preexisting valleys north of the *Apollo 15* landing site (Fig. 3.6).

Mare Tectonism: Deforming the Surface of the Moon

Tectonism is the process whereby planetary surfaces are broken
up, deformed, warped, and stretched. On Earth, a very dynamic
and active planet, tectonism is complex and continual. Lunar
tectonism is very simple, and there are only two principal
classes of tectonic feature. Their mode of formation is well understood, and their distribution patterns are understandable in
terms of the filling of the basins by lava. Moreover, the tectonism
appears to have been confined mostly to a very narrow interval
of time about 3 billion years ago, and subsequent activity has
been minor, with some conspicuous exceptions.

Tectonic features can result either from compression (squeez-
ing the crust) or from extension (stretching the crust). Compression produces a common landform in the maria, the *wrinkle ridge*
(Figs. 2.12, 2.13). Although wrinkle ridges can be very complex in
detail, they are simple in general and are the result of the surface
being squeezed laterally. Think of wrinkle ridges as similar to the
folds that appear in a tablecloth when the edges are pushed together toward the center of a table. Extension results from stretching the crust, and the resulting landform is a fracture or fissure. If
relative movement occurs along the plane of the fracture, it is
called a *fault*. Faults are very common tectonic features on all of
the solid planets and are often found in the maria of the Moon.
Two faults running parallel to each other with a block dropped
downward between them is called a *graben*, the most common
extensional tectonic feature on the Moon (Fig. 5.9). Think of faults
and graben as similar to cracks that develop in a painted surface

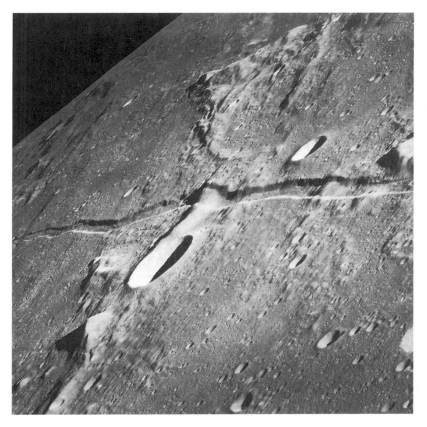

Figure 5.9. A graben, a tectonic feature caused by two parallel, normal faults, with a down-dropped block between them. Most graben on the Moon are found around the edges of the mascon mare basins.

that is bent or bowed outward, where the paint surface breaks because of the strain.

The pattern of deformation of the crust is quite simple in concept. The circular mare basins fill with lava over an extended period of time, up to several hundred million years. As the massive weight of the lava accumulates, it loads the crust, which must accommodate this added weight. The center of the basin sags inward, creating a regional stress field in which the interior of the basin is under compression, forming wrinkle ridges, while the edges of the basin experience tension, creating faults and grabens. This relation is beautifully illustrated by the patterns of deformation around the Humorum and Serenitatis basins, where we find arcuate grabens along the basin edge and circular wrinkle ridges inside the basin (Fig. 5.10). In a nutshell, this is the tectonic story of the Moon. Nearly all the grabens and wrinkle ridges correspond to this pattern of deformation, following the regional trends created by the impact basins.

The History of Volcanism on the Moon

The earliest extrusions of lava on the Moon may have been the outpouring of liquid rock onto the still-cooling, crusted-over surface of the early Moon. Indeed, the line between volcanism and crustal formation was probably indistinguishable in early lunar history. The oldest unequivocal volcanism on the Moon is represented by tiny chips of mare basalt from the *Apollo 14* highland breccias. These fragments represent pieces of a lava flow extruded onto the surface 4.2 billion years ago, a time so remote that we can only guess at what conditions were like on Earth. Volcanic eruptions probably were more or less continuous throughout the period of the heavy bombardment between 4.3 and 3.8 billion years ago. Traces of this epoch of volcanism on the Moon can be found in the tiny fragments of lava in the highland breccias but may also be evident as a chemical signature in cratered terrains. Some regions of the highlands appear to contain relatively large amounts of iron (Plate 9). This iron could represent unsampled highland rocks, but it could also signify flows of iron-rich mare basalt that have been ground up into the regolith of the highlands by the intense impact bombardment of early lunar history.

During the final phases of the heavy bombardment, 3.9 to 3.8 billion years ago, several large, well-preserved basins formed. These basins still have recognizable ejecta blankets, and the smooth, far edges of their debris layers appear to fill craters and other depressions in several areas. Such light plains have all the compositional properties of highland rocks, and Apollo results tell us that the light plains of the highlands are an impact formation, associated with the large multiring basins. However, some

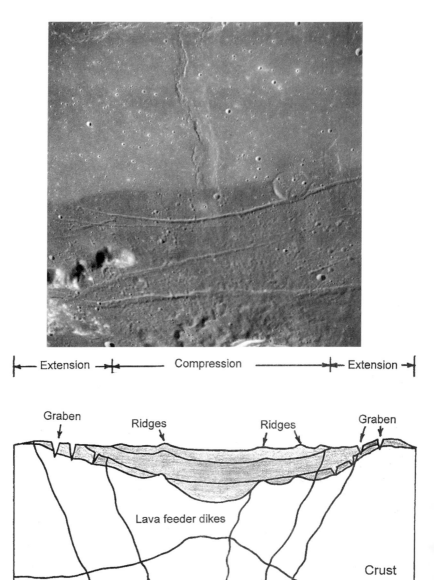

Figure 5.10. The basic tectonic scheme of the Moon. Top: the interior of Mare Serenitatis shows compressional wrinkle ridges while the edge shows extensional graben. In the cross-section drawing (bottom), the loading of the crust by the thick stack of dense lava results in such a stress pattern. After Heiken, Vaniman, and French, *Lunar Sourcebook*, Fig. 4.29.

of these plains display small (1–3 km diameter) impact craters whose ejecta are relatively dark (Fig. 5.11). Be careful not to confuse these dark-halo impact craters with the dark-halo cinder cones found along crater floor fractures (Fig. 5.4). Dark-halo impact craters are found in the light plains of the highlands and cluster in regional groups, such as in the Schiller-Schickard impact basin (Fig. 5.11). Spectral observations of these craters indicate that their low albedo is caused by mare basalt lava making up the crater ejecta—yet this is an area of impact-made plains. How can this be?

The dark-halo impact craters are excavating *buried* deposits of mare basalt (Fig. 5.11). Because the plains that bury the lava flows are themselves 3.8 billion years old, the basalt flows that they cover must be older than this. Thus a map that shows the light plains of the Moon and that displays dark-halo impact craters is a map of ancient maria—basaltic lavas that were emplaced before 3.8 billion years ago. This remote-sensing evidence for the regional extent of ancient maria complements the sample evidence of tiny fragments of lava in highland breccias and indicates that the early Moon was a planet of active volcanism. The extent of the ancient maria is compatible with a broadly declining rate of lava extrusion with time.

The "main phase" of mare volcanism began when the very high rates of cratering typical of early lunar history declined to the point where the lava flows were no longer destroyed and ground up into powder soon after they were extruded. This decline of the impact flux was very rapid between 3.9 and 3.8 billion years ago, leveling off after 3.8 billion years. From that time onward, the extruded lava flows were bombarded by impact, but the cratering did not destroy the flows. These flows make up the visible maria. The earliest lavas from this period are the high-titanium basalts of Maria Tranquillitatis and Serenitatis, basalts returned by the *Apollo 11* and *17* missions. These flows erupted between 3.8 and 3.6 billion years ago. The complete extent of early high-titanium lavas cannot be determined because they are everywhere partly covered by younger flows. We suspect that they are quite extensive over the near side.

A long period of eruption of lower-titanium basalt followed, from 3.6 billion years to an undetermined time, certainly as late as 3.1 billion years ago but perhaps much later. Some of the lavas

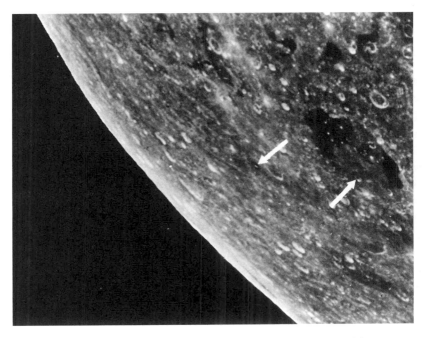

THE FORMATION OF DARK-HALOED IMPACT CRATERS

Formation of crust and early impact cratering

Extrusion of mare basalt

Emplacement of highland material as a result of Orientale event

Excavation of basalts by post-Orientale impact

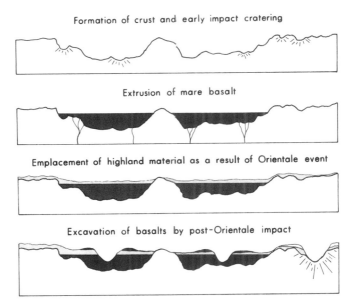

Figure 5.11. Dark-halo craters of impact origin (arrows) indicate ancient mare volcanism. Such craters have excavated basalt from beneath the light-toned, highland plains of the Orientale basin. These lava flows therefore predate the basin, which is 3.8 billion years old. Below: a schematic shows the inferred sequence of events.

from this time were of the high-aluminum variety, particularly in the eastern maria, Crisium and Fecunditatis, as sampled by the Soviet *Luna 16* and *24* sample-return missions. These basalts date from 3.6 to 3.4 billion years and have moderate to extremely low titanium contents. Eruptions of low-titanium lavas in Mare Imbrium at 3.3 billion years ago *(Apollo 15)* and in Oceanus Procellarum at 3.1 billion years *(Apollo 12)* followed. The lavas from the *Apollo 12* site are the youngest mare basalts in the sample collection.

The only other source of information on the age of mare lavas is the relative age data provided by geological mapping. This mapping shows that many different kinds of flows spilled out onto the maria between the dated and sampled eruptions, providing for a continuous filling of the basins from 3.8 billion years ago onward. A key piece of information for lunar volcanic history is one we do not have: What is the age of the *youngest* mare basalt eruption on the Moon? This question has an important bearing on the thermal history of the Moon; if we can discover the age of the last eruption, we will know when the Moon "shut down" thermally, at least to the point where no lava could get out of its interior.

In many books, one reads that volcanism on the Moon stopped about 3 billion years ago. Such a statement is totally without foundation. Many areas of the maria have been identified as having a lower density of impact craters than the sampled flows of the *Apollo 12* site (aged 3.1 billion years). The very young lava flows with well-developed scarps in Mare Imbrium (Fig. 5.1) appear to have high contents of titanium and thorium (a radioactive element) and crater densities of a factor about two to three lower than the lavas of the *Apollo 12* site. Thus these Imbrium flows may be 1.5 to 2 billion years old; we cannot estimate their absolute age any more precisely than this.

Other relatively young flows are found around the Moon, from Oceanus Procellarum in the west to Mare Smythii in the east. One notable example is the *Surveyor 1* landing site (the site of the very first American landing on the Moon back in 1966) within the lava-filled crater Flamsteed P in Oceanus Procellarum. Crater density suggests that these lavas are some of the youngest flows on the Moon, having an age possibly as low as 1 billion years old. Interestingly, television images returned by

Surveyor 1 show that this site has the thinnest regolith of all mare sites visited, a depth estimated to be between 1 to 1.5 m thick. For comparison, the regolith at the *Apollo 12* site (which is underlaid by the youngest *sampled* mare basalts) is about 4 m thick. As already mentioned, regolith thickness reflects the age of the bedrock upon which it forms. Thus we have an independent way to estimate the relative age of the Flamsteed P basalts from the Surveyor data. These images are consistent with an age for these lavas of about 1 billion years old.

Consider for a moment a great "what if?" in the history of lunar exploration. The dogma that volcanism died on the Moon at 3 billion years was a direct consequence of the study of samples returned by the *Apollo 12* mission, and as we have seen, these lavas are about 3.1 billion years old. The *Apollo 12* site was picked near a Surveyor landing craft to demonstrate the new concept of a pinpoint landing, a capability that was essential if Apollo was to visit the hazardous areas of the highlands. Mission planners for *Apollo 12* picked the *Surveyor 3* spacecraft as the target site to test pinpoint landing. What if, instead, they had chosen the *Surveyor 1* site, near Flamsteed P? The samples from *Apollo 12* would have been found to be only 1 billion years old. Instead of a Moon that had geologically "died" over 3 billion years ago, we would have marveled that such a small planet could be volcanically active for almost *4 billion* years. What a startlingly different picture of lunar evolution would have emerged! As it is, we do not know (and will not until we return to the Moon) how long the Moon was erupting lava and when its thermal engine shut down.

The very last gasps of volcanism on the Moon may have occurred about 800 million years ago. Ejecta from the crater Lichtenberg (about 20 km diameter) is covered by a lava flow, which is therefore younger than the crater (Fig. 5.12). Lichtenberg has rays and is a member of the youngest class of craters on the Moon. Estimates of the absolute age of Lichtenberg are uncertain, but craters of this size tend to have their ray systems completely destroyed on timescales of 500 to 1,000 million years. Thus Lichtenberg, along with its overlying lava flow, is probably *younger* than 1 billion years. This mare unit is the youngest lava flow currently recognized on the Moon.

The sum of the evidence suggests that the Moon has been vol-

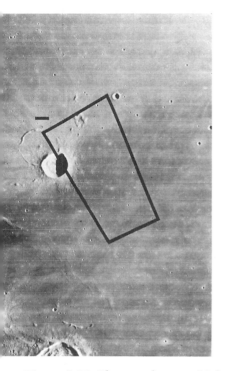

Figure 5.12. The rayed crater Lichtenberg (20 km diameter), partly covered by lava. These lavas must postdate the crater. Because Lichtenberg still shows rays, it must be young in absolute terms, probably younger than 1 billion years. These mare lavas are thus among the youngest basalts recognized on the Moon.

canically active for most of its history. Massive extrusions of lava probably began in dim antiquity, during the era when the Moon's crust was being ground to a pulp by very high rates of bombardment. Extrusion of basalt continued throughout the era of basin formation, including the eruption of the massive amounts of lava that now underlie the highland plains deposited by the Orientale basin (Fig. 5.11). As basins ceased to form, large expanses of mare deposits began to be preserved, forming the complex series of overlapping lava flows and ash deposits that make up the visible maria. Slow, prolonged filling of the near-side basins continued for several hundred million years, gradually loading the crust and deforming the filled basins by interior compression and exterior extension (Fig. 5.10). Some impact craters underwent volcanic modification, including interior flooding and the formation of cinder cones on their floors (Fig. 5.4). Large-scale regional deposits of ash were erupted along the margins of some maria (Fig. 2.16). Younger lava deposits tend to be less voluminous than the older deposits, indicating that the intensity of volcanic activity has declined with time. The youngest lava flows are confined to Procellarum and Smythii (Fig. 5.12). Some of these young eruptions may have occurred since 1 billion years ago, a time that seems remotely old but, in fact, is "only yesterday" in the ancient and silent world of the Moon.

Chapter 6

The Terrae
Formation and Evolution of the Crust

The *terrae* (or highlands) of the Moon are the oldest exposed parts of the original crust (Fig. 6.1). This overwhelming landscape of craters upon craters attests to the lunar history of impact bombardment—the crushing, grinding, melting, and mixing that the crust has experienced. Spectacular, multiring basins (the largest impact craters of the Moon) cover large areas of the highland crust, excavating to great depths and mixing crustal materials on scales of tens of kilometers. Strange landforms that were once puzzling can now be explained as products of this basin formation. To understand this story, we must first read through the overprint of impact. Smooth plains cover vast tracts, some of which are ancient maria that are buried by the debris blankets of the large basins. By looking at the highlands, we look back to the earliest phases of lunar history—to the formation of its crust, probably by large-scale, planetwide melting.

Breccias: Crushed, Fused, and Broken Bits of the Crust

As mentioned previously, a breccia is an aggregate rock made up of bits of preexisting rocks. We have already looked at breccias in relation to the regolith, and their presence attests to the impact bombardment that created the unit in which they are found. In a similar manner, the breccias of the highlands are mute testimony to the epoch of the nearly inconceivable violence that led to the formation of the cratered highlands. Virtually all of the samples returned from the highlands are breccias (Fig. 6.2). Six of the seven meteorites that come from the Moon are breccias. Indeed, because the entire outer layer of the Moon

Figure 6.1. The rugged, cratered highlands of the far side of the Moon. Note the crater-upon-crater nature of the terrain. The complex crater with the elongate central peak is King (5° N, 120.5° E; 77 km diameter).

consists of rocks broken up and partly refused by impact, one could argue that this layer is one gigantic breccia.

We typically classify breccias by the types of rock fragments *(clasts)* that they contain, by the nature of the breccia ground-mass *(matrix)* that holds the fragments together, and by their chemical composition. Regolith breccias are essentially fused chunks of lunar soil. They can be recognized by their glass-rich matrices, some of which may be partly recrystallized, and by the presence of tiny shards of agglutinates that occur as clasts; only regolith breccias have such fragments. In addition, regolith breccias contain solar wind gases, such as hydrogen, helium, and the rare gases argon, krypton, and neon. Clasts consist of the rock types found in the regolith in which they formed. In mare

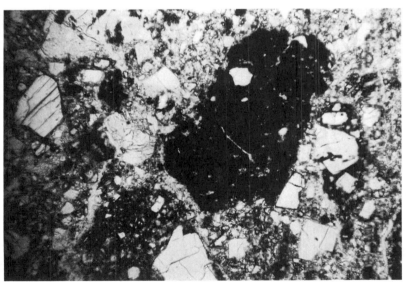

Figure 6.2. A breccia from the highlands. This rock, sample 67015 from the *Apollo 16* Descartes site, is made up of myriad fragments of earlier, broken rocks, some of which are breccias themselves. The top view is the hand specimen; the bottom view is a microscopic thin-section view (see Fig. 5.2).

regolith breccias, the clasts consist of mare basalts and their minerals, agglutinates, fragments of older regolith breccias, and small bits of highland rocks. Highland regolith breccias consist of chips of highland rocks, crushed mineral fragments derived from highland rocks, agglutinates, and, very rarely, tiny bits of mare basalt and their minerals. Regolith breccias from the highlands are noticeably lighter than their mare counterparts, a reflection of the generally lower iron content (and higher aluminum content) of the highland regolith compared with the maria.

The most important type of rock from the terrae is the highland breccia, which makes up the rugged, cratered terrain of the Moon. Highland breccias come in a variety of types, including breccias of a single rock type (made by an impact into a terrain made up of one rock type), two-rock mixtures, and mixtures of multiple rock types. Often highland breccias contain clasts of older breccias, and sometimes this breccia-within-breccia texture can extend to four or five generations of breccia formation. The sizes of the clasts in the highland breccias are highly variable, ranging from submicroscopic ones to large clasts as big as a house ("House Rock" was visited and photographed [Fig. 3.9] during the *Apollo 16* mission to the Moon). At hand-specimen scales, a breccia typically contains clasts ranging in size from a few centimeters to microscopic (Fig. 6.2).

The grain sizes of clasts within highland breccias follow a *power-law* distribution, similar to that of the size distribution of craters on the Moon. Essentially, this means that small grain sizes dominate and that increasingly larger clasts become increasingly rare. In large-scale systems that crystallize from magma, the contact relations between rock types are very important because they contain clues to the original conditions of crustal formation. Because the rocks of the crust have been crushed by impacts, it is unlikely that we can find evidence for igneous contacts between rock types. The pulverization of the Moon into breccias greatly complicates our reading of the origin of the original crust while at the same time it provides an abundance of information about the cratering process, particularly at the larger scales of basin-forming events.

Breccias can be fragmental, impact-melted, or granulitic. Fragmental breccias are basically *sedimentary* rocks, aggregates of a variety of older rock types bonded together within the ejecta blanket of an impact crater (Fig. 6.2). Impact-melt breccias are *igneous* rocks, created from shock-melted, liquid rock that contains clasts of minerals and other rocks, all bonded together into a single, new rock type. Granulitic breccias are *metamorphic* rocks, formed by recrystallization of older grains without melting. Lunar granulites are found at all sites on the Moon, have a composition that closely resembles the bulk composition of the crust, contain no KREEP component (discussed later in this chapter), and appear to have formed before about 4.1 billion years ago. Granulites underwent slow recrystallization under very high temperature conditions (called *annealing*) and are direct evidence for high crustal temperatures for at least the first 500 million years of lunar history.

Impact Melting and the Formation of Breccias

The collision of a large meteorite with a planet has catastrophic consequences. The energy contained by the speeding object is instantly transferred to the surface of the planet. This energy is divided into fractions: Some of it breaks up and crushes the target, ejecting material out of the crater onto the surrounding terrain, and some of it melts and vaporizes the target and impactor. Impact (or shock) melting of the crust is an important process in the creation of the cratered terrain of the Moon. Many of the samples returned from the highlands are impact-melt breccias of various types (Fig. 6.3). They offer clues to the composition of the crust as well as to the formation of large craters and basins.

Impact-melt breccias are aggregates of a variety of different highland rock clasts, bound together by an igneous (once molten) rock matrix. The matrix can have a wide range of grain sizes, from so small that the individual grains cannot be seen even with the most powerful microscopes to relatively coarse, comparable to the crystals seen in the mare basalt lavas. Some of these impact-melt rocks resemble lava samples so much that some workers who conducted the initial studies of them thought the rocks were volcanic lavas from the highlands.

Textures seen in impact melts attest to a molten state at the

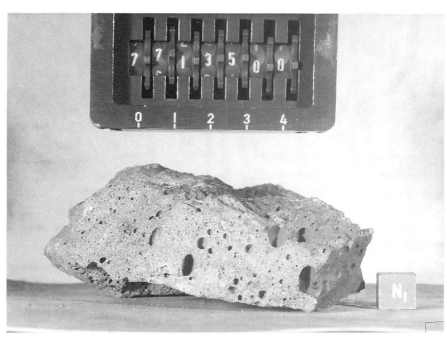

time of their formation (Fig. 6.3). Melts that are relatively free of clasts tend to have a coarser matrix and closely resemble samples of lava. The finer-grained rocks, including those with submicroscopic grains, tend to be much richer in clasts. These clasts are almost entirely fragments of minerals, mostly plagioclase, the dominant mineral in the highlands. The sizes of the clasts contained within the impact melts range downward to the sizes of their matrix grains, making it very difficult to distinguish between matrix and clast in many of the impact-melt samples. Because grain size in igneous rocks reflects the length of time they take to cool, the fact that the finest-grained rocks contain the most clasts suggests that the clasts "chilled" the very hot shock melt. The more clasts entrained into the melt as it spread along the crater floor, the more rapidly the melt cooled. Conversely, coarse-grained impact melts must have cooled more slowly, probably within the insulated interior of a sheet of impact melt.

In determining the crater of origin, the texture of impact melts is less important than their chemical makeup. The composition of impact melts tells the geologist about the impact target. From the study of craters on Earth, we have found that the chemical composition of impact melts represents an average of all of the rock types that make up the target. Some of the terrains that served as targets for the largest terrestrial craters were quite diverse and complex, with rock types ranging from dark, iron-rich lava to light, silica-rich granite rocks. Curiously, the shock melt created from a large impact into such a terrain has a single, very uniform composition. Careful work has documented that this composition can be represented as a physical mixture of the target rocks in some proportion.

This experience seems to defy common sense. How can a very large, growing crater, encountering dark, iron-rich rocks at one place and granites at another, produce a melt sheet of extreme

---

Figure 6.3 (opposite). An impact-melt breccia from the rim of the Serenitatis basin, the *Apollo 17* landing site. The rock is made up of many small mineral and rock fragments, cemented by a fine-grained matrix that crystallized from a melt. Because of similarities to melt rocks from impact craters on Earth and because of its old age (3.87 billion years), we believe that this rock was created in a very large impact, probably a basin-forming event.

uniformity? Yet some of the very largest craters in Canada display this melt homogenization, including those craters with diameters of over 100 km. In these impact craters on Earth, we know the bedrock geology of the crater target as well as the contact relations between the impact melt sheet and the crater floor. On the Moon, neither of these pieces of information is generally available.

Lunar impact melts tend to be chemically homogeneous within a given rock sample or even a large boulder. In addition, multiple samples of melt from single, or even different, Apollo landing sites appear to have the same, or very similar, compositions. Does this indicate that the samples come from the same impact? Or does it show that the melts formed from a number of impacts into a terrain of identical composition? The serious consequences of the lack of geological control for most of the samples from the Moon now become evident. Scientists typically extrapolate results from study of terrestrial impact melts to lunar rocks. Indeed, our fundamental understanding of the process of impact melting comes from the use of craters on Earth as a guide. However, this guide must be used with caution. No melt sheet on the Moon has been sampled in place, and all samples come from the regolith. Thus the number of impacts and the original composition of their target(s) remain unknown. All we have is the testimony of the rocks themselves.

One of the principal mysteries of impact melts on the Moon is the chemical composition of their matrices, which make up the bulk of the volume (typically more than 90 percent) of these breccias. Nearly all lunar impact melts have a composition known as LKFM, meaning "low-K Fra Mauro" basalt and named after the landing site *(Apollo 14)* where they were first found. This composition is roughly the same as that of basalt, but keep in mind that these rocks are *not* of volcanic origin but are *impact* melts. The puzzle is this: If all the impact melts have this LKFM composition, where are the *targets* in which they formed? The average composition of the crust (determined in a variety of ways, including remote sensing and direct sampling) is much richer in aluminum, poorer in iron and magnesium, and nearly completely lacking in the trace elements, including the radioactive elements uranium and thorium. The LKFM composition of the impact melts from the Moon is nearly the exact reverse of the composition of the average highlands. How can an impact melt be made that has a composition completely different from that of the target rocks in which the craters formed? Are we wrong about how impact melts are created?

A solution to this problem might lie in the scale of impact events on the Moon compared with those on Earth. Craters with preserved sheets of impact melt on Earth, such as the Mani-couagan crater (55 km diameter) in Quebec, are between 50 and 100 km in diameter. Multiring basins on the Moon are at least 10 times bigger; the Imbrium basin is over 1,100 km in diameter, larger across than the state of Texas. As such, the largest impacts on the Moon dig into and excavate many kilometers into the crust. Impact melts from a basin-forming event probably are generated at great depths, possibly involving the lower crust and upper mantle. Because the ejecta of the largest basins (which excavated deep into the Moon) show an iron-rich and aluminum-poor composition compared with that of the surface of the highlands, the composition of the lower crust is believed to be different from that of the upper crust. In fact, this iron-rich composition is quite similar to LKFM. So it is likely that LKFM impact melts are the products of the melting not of the upper crust but of the lower crust (and possibly the upper mantle) of the Moon.

If these impact melts, abundant in the Apollo collections, are created during the very largest impacts, those that formed the basins, they are our best samples of the lower levels of the crust. They probably formed by melting crustal levels 20–60 km deep (Fig. 6.4). The variations in composition of the LKFM melts could represent slightly different levels of sampling, with the more aluminous melts (from *Apollo 16*, near the Nectaris basin) coming from shallower levels of the crust and with the more mafic melts (the *Apollo 15* black-and-white rocks near the Imbrium basin) coming from deeper in the Moon. Such different depths also reflect the size of the respective impacts; the Nectaris basin is about 850 km in diameter, whereas the Imbrium basin is over 1,100 km in diameter. Thus Imbrium, being larger than Nectaris, would dig down deeper into the Moon, melting more iron-rich, aluminum-poor zones of the crust.

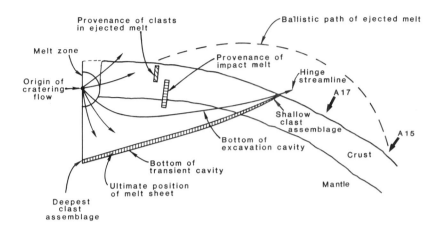

Figure 6.4. Diagram showing a section through a basin excavation cavity. The impact melt is formed by very high peak shock pressures near the point of impact. As the cavity grows, the melt is mixed (accounting for its relative homogeneity) and lines the floor of the crater. This melt sheet eventually covers the final floor of the basin, having mixed together many fragments of crushed rock and mineral debris.

The clasts contained within these LKFM melts consist entirely of fresh, crustal rocks and their associated mineral debris. They show no regolith agglutinates or glasses as clast types and have no solar wind gases. The rock clasts in LKFM melts appear to display the complete variety of crustal rock types; indeed the presence of such clasts is our principal way of sampling these important rock types. As discussed in the last chapter, rare clasts of mare basalt are also found and represent remnants of ancient mare volcanism in the highlands. One curious fact about lunar melt breccias is that the chemical composition of the matrix (the LKFM composition) *cannot* be made by the fusion of the observed rock and mineral clasts, which tend to be more aluminous and less rich in iron and magnesium (less mafic). If the supposition that these rocks come from basin-forming impacts is correct, then the shock melt of the basin impact comes from deep within the Moon, and the clastic debris included in the melt must come from different, probably shallower, levels of the crust.

Cataclysmic Violence?: The Early Cratering History of the Moon

One aspect of impact-melt rocks is extremely important to our reconstruction of lunar history: These rocks are the only impact-produced samples that are suitable for age measurements. This is because melted samples completely reset their isotopic clocks, and age data from these samples represent the age of crystallization (cooling) of the rocks. Such an event occurs on relatively short timescales after an impact (on the order of hundreds to thousands of years for even the largest impacts, much shorter time spans than the ages of these features). So when a melt breccia is dated, the age obtained reflects the age of some impact. If a given sample can be plausibly associated with a specific impact, either because of the proximity of its collection site to the feature or some other aspect of its geology, these data for rock ages can be used to assign *absolute ages* to the large impact events of lunar history. This task was one of the prime scientific objectives of the Apollo missions.

Obtaining useful absolute ages from rocks is a difficult business. There are various techniques used to get these ages, and all are based on measuring the minute amounts of radiogenic isotopes in the rocks and calculating how much of these isotopes has decayed into another element. Because this decay occurs at a known rate, the age of the rock can be computed. The technique assumes that the isotopic system of the rock has not been disturbed since it crystallized. Sometimes it is very difficult to obtain a satisfactory age because the clasts in an impact melt can release minute amounts of radiogenic gas, altering the apparent age of the rocks. Some samples are so clast-rich or have had their systems so disturbed that satisfactory ages cannot be obtained at all. Almost all of the ages obtained for impact melts from the Moon are determined by the argon technique, which involves irradiating the sample in a nuclear reactor, heating the sample to high temperatures, and measuring the tiny amounts of different types of radiogenic argon released from the sample. This technique is used because impact-melt breccias from the Moon have very small grain sizes and because this method does not require the separation of individual minerals from the sample, as is commonly done with coarse-grained rocks.

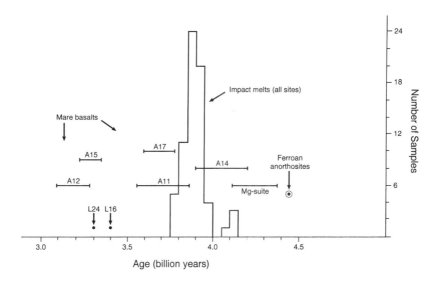

Figure 6.5. A histogram that shows the occurrence of ages in the highland impact-melt rocks from the Apollo samples. The black bars show the range of measured ages for many samples of basalt and highland igneous rocks from various sites on the Moon; the three dots (L16, L24, and ferroan anorthosites) indicate the measured ages of single samples. Note that most of the impact-melt breccias seem to have formed between 3.8 and 3.9 billion years ago. Does this mean that an impact cataclysm occurred on the Moon or that we sampled only one or two large events? We need to return to the Moon to answer this question.

The astonishing result from the dating of the lunar samples is that the impact-melt breccias all appear to have formed at very nearly the same time: between 3.8 and 3.9 billion years ago (Fig. 6.5), a duration of time that corresponds to less than 3 percent of the history of the Moon! This result is obtained from dating the samples from all of the Apollo sites, which are spread across the near side and are separated in time (being on different geological units) as well as by space. Closer examination of the age data reveals some small differences. The impact-melt breccias from the *Apollo 16* site form groups by chemical composition and by age. The most aluminous melt group is the youngest, at 3.82 billion years old, whereas the most abundant melt group, an aluminum-rich variety of LKFM, is about 3.92 billion years old.

Data such as these suggest that several events may be represented, but the impacts are very closely spaced in time. Ages from the *Apollo 17* site cluster very closely around 3.87 billion years; this fact, in addition to the site's proximity to the edge of the Serenitatis basin, leads us to believe that this age reflects the age of the impact that formed the Serenitatis basin.

One of the most important ages to determine is the age of the Imbrium basin. This giant feature is used to subdivide geological time on the Moon (see Chapter 2). Two Apollo missions were sent to sites specifically chosen to sample deposits of the Imbrium basin: *Apollo 14* at Fra Mauro (Fig. 3.4) and *Apollo 15* at the base of the Apennine Mountains (Fig. 3.6). *Apollo 14* probably returned material ejected from the Imbrium basin, although it is not certain which of those samples are basin related. The *Apollo 14* site is far from the rim of the basin, and it is uncertain whether or how much impact melt (the product needed for dating) is ejected such distances. At the *Apollo 15* site, such impact melt is probably present, but the best candidates for melt from the Imbrium basin have poorly defined ages. These rocks appear to be about 3.84 billion years old. Thus the Imbrium basin, the most important feature on the Moon to date, does not have a well-determined age; an age of about 3.84–3.85 billion years is the current best estimate.

All of the known ages of lunar impact melts are within 80 to 100 million years of each other (Fig. 6.5). Does this narrow time interval mean that many different impacts all occurred at about the same time, or does it signify that only one or two impacts provided samples at many different sites? This question is at the heart of the debate about early lunar history. Devotees of the former view hold that the early Moon underwent an intense, distinct period of impact bombardment, or a *cataclysm*, about 3.8 billion years ago. This cataclysm would have produced not only the large basins but also the multitude of large, overlapping craters in the highlands. Scientists who dispute the cataclysm contend that the Apollo missions visited sites dominated by deposits of only the youngest large basins, predominantly the Imbrium, Serenitatis, and possibly, Nectaris basins.

Which view is correct? This problem has far-ranging consequences because the evidence for early bombardment on the

Moon is used to interpret the early histories of all the terrestrial planets. In favor of the cataclysm viewpoint is the testimony of the rocks themselves. All of these ages of 3.8–3.9 billion years come from impact-melt breccias from the highlands; thus these ages are dating impact events. It has been argued that older ages are not present because the high rates of bombardment in early lunar history have "reset" the ages of earlier rocks, preventing older ages from being preserved Yet such an effect is unlikely because (1) it is nearly impossible to reset a rock age, short of completely remelting the rock, and (2) the existence of ancient mare basalts that are not remelted (some as old as 4.2 billion years, see Chapter 5) shows that very old rocks *have* survived this period. Thus the melt rocks probably represent new impacts, and their peak in ages (Fig. 6.5) probably reflects a real cratering record. The question is, how many impacts are represented by these rocks? If the ejecta blanket of a basin is made up of the ground-up melt sheets of many different highland craters, then the ages of the returned melt samples reflect the ages of formation of many, if not most, of the highland craters. Not only all of the visible basins but also all of the large craters would have formed around this time (because the large craters are known, by relative age assignments, to be sandwiched in time between the major basins).

The argument against the cataclysm suggests instead that the peak in melt ages actually reflects the influence of only a few large impacts. In support of this view is the composition of the melts, the vast bulk being LKFM (meaning lower crustal), which suggests that they formed in basin impacts. Because the Apollo missions landed within the geological influence of only three large basins (Imbrium, Serenitatis, and Nectaris, see Fig. 2.11) and because the melt breccia ages tend to cluster by composition at each site, these ages would reflect the ages of only these three basins. The Nectaris basin would be the oldest at 3.92 billion years, its melt represented by the *Apollo 16* aluminum-rich LKFM melt. The Serenitatis basin would be the next oldest, represented by the *Apollo 17* melt samples; its age is 3.87 billion years old. Imbrium, as mentioned above, is likely to be around 3.84 billion years old, dated by the *Apollo 15* impact-melt samples.

This assignment does not completely disprove the concept of

the cataclysm because the Nectaris basin is still relatively young; an age of 3.92 billion years for Nectaris requires a sharp increase, or *spike,* in the early cratering rate. One could consider this a cataclysm of sorts. In fact there is a spectrum of possibilities, ranging from a cataclysm responsible for all of the craters on the Moon to a minor and temporary peak in a generally declining cratering rate. Virtually any position within these broad bounds can be defended scientifically.

We must return to the Moon to solve this problem. The easiest way would be to explore the mountainous rim of the largest known impact crater on the Moon, the South Pole–Aitken basin. From its very high density of impact craters, we believe that this feature is the oldest basin on the Moon (Plate 10). What we do not know is its absolute age. A mission or series of missions to the mountains of this basin (Fig. 6.6) could search for and sample impact-melt breccias produced in the basin-forming impact. If this basin is 3.9 billion years old, then there *was* a cataclysm, in its most extreme form. Any age for South Pole–Aitken older than this argues for a less severe cataclysm. If the basin is around 4.3 billion years old, there is no need for a cataclysm, except to explain the seemingly young 3.92-billion-year age of the Nectaris basin. A sample-return mission to massifs of the South Pole–Aitken basin could address many other scientific questions as well; because this basin is large enough to have excavated the complete crust, there is also the intriguing possibility of returning samples from the mantle.

Is it possible to test the idea of a cataclysm without a return to the Moon? Although results would not be definitive, two things could be done while we wait for a return to the Moon. We could measure the age of very ancient mare basalts by the argon method (used to date impact melts). If the ages obtained by this method match the results obtained by the separation-of-minerals method, used on the old lavas, it would demonstrate that the argon clock is valid (it has not been reset by, for example, the high internal temperatures within the early Moon) and therefore that the age data for melt breccias reflect the ages of impacts. A second test would be to measure the ages of highland melt rocks that occur as tiny clasts in the lunar meteorites (the majority of which are regolith breccias); such meteorites could come from

Figure 6.6. The mountain rim of the South Pole–Aitken basin (see Plate 10). A future mission to these hills could get samples of the impact melt produced during the creation of the largest crater in the solar system. From such rocks we not only would learn much about the crustal composition but also could resolve the question of whether or not a cataclysm occurred.

nearly anywhere on the Moon. If these melts formed at the time of the cataclysm (3.8 billion years ago), this would argue that the high-impact flux was a Moon-wide phenomenon and that we are not looking at a local effect produced by the selection of the Apollo sites around the central near side.

This is an important problem to solve. If there was a cataclysm, even a small one, how did it work? Why would a planet undergo an intense bombardment over 600 million years after it had already accreted into a single body? One possibility is that a large planetoid, or "submoon," wandered into the Earth-Moon system, was broken up into rubble by tidal forces, and formed a cloud of debris that rained onto the early Moon (and possibly Earth as well). If such a scenario is correct, the lunar rock record doesn't really tell us how to interpret the early history of the other planets—each object could have its own, unique bombardment history. On the other hand, if we can show that there was no cataclysm, the record from the lunar samples can help us interpret the early histories of the other terrestrial planets. Either way, we need to know the answer to this question.

Plutonic Rocks: The Crust of the Original Moon Pie

Very early in the study of lunar rocks, we realized that the breccias of the highlands are made up of pieces of much older rocks. These rocks look like the rocks on Earth that form at great depths and cool very slowly. A rock that forms by the very slow crystallization of a magma (liquid rock) at depth is called a *plutonic* rock. On Earth the granite that dominates the crust of the continents is an example of a plutonic rock. Granite forms by the slow cooling of a magma rich in silicon and aluminum. It is often found in the cores of mountain ranges, geological terrain that has been thrust up by powerful tectonic forces, exposing rocks formed at depth. Granites may cool on timescales of thousands to hundreds of thousands of years. This slow cooling allows crystals to grow very large, accounting for the coarse-grained nature of plutonic rocks.

During study of the *Apollo 11* regolith, we recognized small bits of white rock among the generally dark, mare basaltic moondust. It was proposed that these samples came from the highlands, and from this hypothesis, it was imagined that the highlands are composed of the plutonic rock *anorthosite*. Anorthosite is a rock made up almost entirely of one mineral, plagioclase (Plate 11). Most rocks (of any type) are made up of at least two minerals and usually of three to four major and a host of minor minerals. The idea that the crust is rich in plagioclase is supported by other observations, including the light tone (high albedo) of the highlands compared with the dark maria and an analysis of the soil by *Surveyor 7* on the rim of the crater Tycho (Plate 2, Fig. 3.1) in 1968, before Apollo landed on the Moon. This chemical analysis revealed a relatively high concentration of aluminum, consistent with anorthosite, and in fact this possibility was cited (along with several other possibilities) by the Surveyor investigators. The sample returns from the highlands by the later Apollo missions confirmed the highland anorthositic composition earlier inferred from the *Apollo 11* soils.

The anorthosites contain small amounts (much less than 5 percent) of the minerals pyroxene and olivine (Plate 11), both of which tend to be rich in iron relative to magnesium. From this property, scientists sometimes refer to the anorthosites as *ferroan anorthosite* (from *ferrum*, Latin for "iron"). Because anorthosites are almost entirely one mineral, they are very difficult to date with the mineral-separation techniques we use to date most lunar samples. One ferroan anorthosite from the *Apollo 16* site does contain enough mafic minerals to separate, however, and it was dated successfully at 4.42 billion years old. We know from the relationship of isotopes in other dated samples that the initial ratio of strontium, one of the elements used in the rubidium-strontium dating method, gives an indication of the crystallization age of a related sequence of rocks. This indirect evidence suggests that all of the ferroan anorthosites are very old— possibly as old as the Moon itself (4.5 billion years, Fig. 6.5).

No magmas or lavas with the chemical composition of anorthosite have ever been recognized. Anorthosite contains about 35 percent aluminum oxide by weight, the greatest amount of all rocks. It has been known for some time that certain rocks formed from magma have done so by a process whereby minerals are *removed* from the molten system as they crystallize. For anorthosites, plagioclase crystallizing from the melt must be segregated somehow and then accumulated in one place to form anorthosite. How can crystals be removed from a body of liquid rock? Gravity can be responsible: If a crystal is denser than the liquid, it will sink; if it is less dense than the liquid, it will float. Plagioclase is a low-density mineral (about 2.9 g/cm^3) and would float in most magmas of lunar composition.

We think that the anorthosites formed by plagioclase flotation in a body of magma. But the anorthosite is global—it occurs everywhere in the highlands. The distribution of iron in the highlands, as revealed by data from the Clementine mission (Plate 9), shows that vast areas of the terrae are made up of anorthosite. The gravity contrast between maria and terrae and the average relief of the highlands suggest that the highland surface is under-laid by at least a 20 km thickness of plagioclase. The implication of such widespread anorthosite is astounding: If anorthosite forms by the floating of plagioclase in a magma, the whole Moon must have been a molten body of magma! This model, the "magma ocean hypothesis," is now widely accepted and, to a first order, is how the earliest crust formed. We will discuss this model further, after examining some of the other rock types that make up the crust.

Other rock types from the highlands are also rich in the mineral plagioclase but have substantial amounts of other minerals as well (Plate 12). These minerals are mainly olivine and pyroxene, iron-and-magnesium-rich silicates that have a crystal structure different from that of plagioclase. In these rocks the olivine and pyroxene make up about one-half of the minerals of the rock. If a rock is made up of plagioclase and olivine, it is called a *troctolite*. If the rock is made up of plagioclase and pyroxene, it is called a *norite*. The troctolite and the norite of the highlands, along with some other varieties of rock, collectively make up the second broad group of highland rocks: the *Mg-suite*, so-called because the abundant olivine and pyroxene of these rocks is rich in the element magnesium (abbreviated Mg in chemical notation). The rocks of the Mg-suite are found at every landing site (at varying abundance), typically as clasts in highland breccias and impact melts.

In contrast to the nearly uniform antiquity of the ferroan anorthosites (all are 4.4 to 4.5 billion years old, as far as we can tell), the Mg-suite rocks have a range of ages, from 4.5 to 4.2 billion years old (Fig. 6.5). This range, in addition to details of the chemical makeup of the rocks, indicates that the Mg-suite did not form in a single body of magma like the anorthosites but rather in a number of different magma bodies. The rocks of the Mg-suite thus reflect a long and complex period of crustal magmatism and formation. Global maps show that the Mg-suite is a minor contributor to the composition of the highland surface. Chemical data obtained from lunar orbit show that Mg-suite rocks underlie less than about 10 percent of the area covered by the Apollo missions (Plate 13). However, several lines of evidence suggest that the Mg-suite may make up a substantial fraction of the *lower* crust. The clasts contained in impact melts from basins are nearly all rocks of the Mg-suite and minerals derived from such rocks. The rims and floors of the largest basins, where many kilometers of crust have been stripped off, expose Mg-suite rock types. These observations suggest that intrusions of the Mg-suite are abundant in the lower crust and, in fact, may make up most of the lower half of the crust.

The last major rock type of the crust is one of the most puzzling in that it was first recognized as a chemical component and was found as a rock type only after considerable study. It was noted that breccias from the *Apollo 12* and *14* landing sites were particularly enriched in a group of the elements that include samarium, uranium, and thorium (making this component radioactive). These elements do not enter into the crystal structure of the common minerals during magma crystallization because of their size or charge and are thus termed "incompatible" trace elements. As crystallization proceeds, these trace elements tend to become concentrated in the remaining liquid phase of the magma. When crystallization is 99 percent complete, the liquid that is left will be extremely rich in incompatible elements. This component has been given the name *KREEP,* an acronym from the letters for potassium (K in chemical notation), the rare earth elements (REE), and phosphorus (P).

Varying amounts of KREEP are present in nearly all of the breccias, especially in the impact melts that make up so much of the Apollo collections. As a chemical component in breccias, KREEP has no mineral makeup. However, some igneous rocks contain substantial amounts of KREEP. The first to be recognized as a new rock type was a collection of small fragments of basalt from the *Apollo 15* landing site (Fig. 6.7). These rocks are volcanic lavas but are more aluminous than mare basalt and contain a significant amount of KREEP. These KREEP basalts, as they are called, were extruded onto the lunar surface about 3.84 billion years ago and are samples of nonmare volcanism on the Moon. Remote-sensing data show that these fragments come from a light plains deposit inside and younger than the Imbrium basin. This unit, the Apennine Bench Formation (Fig. 3.6), is the best example of a KREEP volcanic lava flow on the Moon.

Other crustal rocks have various amounts of KREEP. Another

Figure 6.7. The highland volcanic lava, KREEP basalt. This sample comes from the *Apollo 15* landing site and is derived from the light-plains Apennine Bench Formation (see Fig. 3.6). These small samples are the only ones of highland, nonmare volcanism in the Apollo collections. The relative abundance of such rocks in the crust is unknown.

rock from the *Apollo 15* landing site is composed of plagioclase, pyroxene, and quartz, a mineral made of silica and rare on the Moon. This rock contains KREEP at levels about five times greater than the KREEP basalts and may be related to them. Other small fragments of rare lunar rock types, such as granite (only a few fragments known), are also relatively rich in KREEP. Even some mare basalts contain it, as do some ancient basalts that occur as clasts in *Apollo 17* breccias. KREEP is completely absent from the granulitic breccias, suggesting that whatever process added that component to breccias did not operate on these rocks. The plagioclase-rich granulitic breccias are probably created near the surface or in the upper crust, where KREEP is rare, whereas the lower-crustal melts of LKFM composition incorporated large amounts of KREEP. The occurrence of KREEP in lunar rocks suggests that it resides in the deep crust and is brought to the surface by basin-scale impacts or by volcanism, in which lavas moving through the lower crust can bring the rock up from depth.

Global Melting and Floating
Plagioclase: The Magma Ocean

The large amount of plagioclase (in the form of anorthosite) within the crust is hard to explain by any process short of near global melting and the subsequent separation of plagioclase by flotation. The idea that an ocean of magma existed early in the history of the Moon was immediately attacked as implausible, but this concept has held up for over 25 years, and everything we learn about the Moon supports such a model. In fact the idea that a planetary body as small as the Moon might experience early global melting has been used as a guide to interpret the early crustal formation of the other terrestrial planets.

Granting the existence of a magma ocean, what was this episode like? Contrary to some initial imaginings, the Moon was probably not a glowing sphere of liquid rock in space. Even if the Moon was largely molten at any given time, a chilled crust would have quickly formed as insulation against the cold vacuum of space. This chilled rind would have had a chemical composition identical to that of the bulk of the Moon, and we have searched the sample collection for pieces of it, although none have been identified. Instead of a true "ocean" of magma, think of the magma ocean as the partly liquid, partly solid outer shell of the Moon. The earliest phase of lunar history was the accumulation of many small bodies, a process known as *accretion*. If this happened rapidly enough, so much heat would be liberated by accretion that this heat would not be able to escape and the Moon would melt.

Experiments have shown that whereas plagioclase crystallizing out of this large body of liquid rock would be less dense than the liquid in which it formed and would tend to be buoyant, mafic (i.e., iron- and magnesium-rich) minerals such as olivine and pyroxene are denser than the magma and would sink. In the magma ocean model, sinking mafic minerals ultimately become the mantle and serve later as the source regions for the mare basalt lavas. In such a molten mass, convection (the gravity-driven stirring of a hot system) would be vigorous, allowing the floating plagioclase crystals to reach the surface by riding the hot, upwelling currents in the molten system. These plagioclase masses would coagulate near the surface, forming masses of anorthosite, similar to a "rockberg" in an ocean of magma (Fig. 6.8). The separation of the crust and mantle by the magma ocean was largely complete by 4.4 billion years ago, early in the history of the Moon. This time is inferred both from the ages of the ferroan anorthosites and from the "model" ages of the mare basalts, which are related to the age of their original sources in the mantle, the sunken olivine and pyroxene crystals from the magma ocean.

The formation of the anorthosite crust and the olivine-pyroxene mantle was only the earliest phase of the igneous history of the Moon. Over the next 300–400 million years, many liquid rock bodies formed deep within the Moon, rose up through the very hot mantle and crust, and squeezed themselves into and among the anorthosites. The process whereby existing rock is replaced by intrusions of new igneous rock is called *assimilation*. If magma rich in magnesium encountered large masses of anorthosite, plagioclase would be assimilated (remelted and incorporated into the magma), and the resulting rock mass, when cooled, would form troctolite, one of the Mg-suite rocks. We think

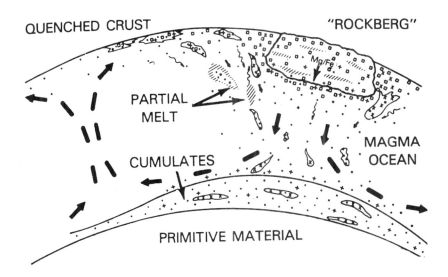

Figure 6.8. Diagram showing the outer portion of the Moon 4.5 billion years ago, when the Moon was covered by a thick layer of liquid rock, the magma ocean. In such a system, low-density minerals (plagioclase) floated, forming the anorthosite crust, while high-density minerals (olivine, pyroxene) sank, forming the mantle. We think that the deep interior of the Moon may be material of unmelted, primitive composition. After J. Longhi, "Pyroxene Stability and the Composition of the Lunar Magma Ocean," *Proceedings of the Ninth Lunar Science Conference* (New York: Pergamon Press, 1978), 285–306.

that most of this intrusive activity by Mg-suite rocks was concentrated near the base of the crust, resulting in a crust that is composed mostly of anorthosites in its top half and predominantly of a complex series of Mg-suite plutons in its lower half (Fig. 6.9).

Along with this extensive intrusive activity by Mg-suite rocks, mare volcanism was active very early in lunar history. The oldest mare basalts, at 4.2 billion years, just overlap in age with the youngest Mg-suite rocks, also 4.2 billion years old. So the last stages of crust-forming magmatism were accompanied by the eruption of floods of basaltic lava. As the crust cooled and solidified, the last dregs of the former ocean of liquid rock became the component we call KREEP. Originally, this radioactive liquid probably resided near the base of the crust, a place where it could be incorporated into the rising magmas that

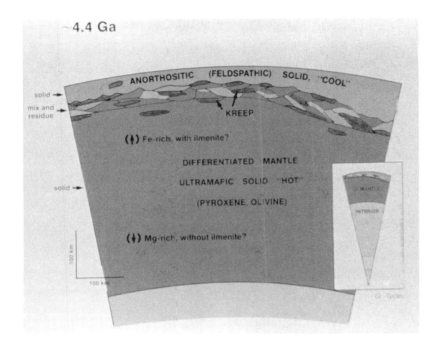

Figure 6.9. Cutaway view of the Moon near the time of completion of crustal formation. The Mg-suite rocks have intruded the original anorthosite crust, making up most of the lowest crustal levels. Convection (the rising of hot currents from the deep interior) would carry magmas up toward the surface, resulting in much crustal magmatism.

intrude the crust. However, its subsequent history (after formation at 4.3 billion years) is complex and involves remelting, assimilation, and impact mixing. KREEP became incorporated into volcanic, plutonic, and impact-generated magmatic liquids. This diverse history accounts for its preponderance and its variability in concentration.

Impacts would have churned and stirred the magma ocean, destroyed fragments of the earliest crust, remelted newly crystallized fragments of the crust, and mixed the lot into a complex, multicomponent breccia. The continued formation of craters and basins excavated and mixed a newly created crust and may have largely erased the original contacts of rock units, greatly complicating our ability to reconstruct lunar history. The biggest basins have melted large volumes of the crust, creating LKFM impact melts and bringing up fragments of the deep crust for our inspection. The creation of the visible craters of the high-lands has thoroughly broken up and churned the upper crust into a massive deposit of breccia, the megaregolith. The high rates of heat flow within the early Moon caused the recrystalliza-tion of some of this early breccia into granulites.

The formation of the megaregolith has mixed together the materials of the crust, with such mixing being in a mostly verti-cal, not lateral, manner. We see preserved on today's Moon some of the original provinces of different composition, provinces cre-ated in the era of crustal formation over 4 billion years ago. Mare basalts were extruded onto this complex, recycled surface, resulting in the creation of the mare-highlands dichotomy we see today. Since then, only the occasional formation of an impact crater has disturbed the silence of this once-violent, turbulent planet, whose earliest events are shrouded by the overprint of an impact record that may hold the key to understanding the early history of all the planets.

Chapter 7

Whence the Moon?

The origin of the Moon has been pondered, modeled, used as a justification for lunar exploration, studied, and endlessly debated. The popular press has often asserted that determining the Moon's origin is the ultimate goal for scientists. As we have seen, much of the story of the Moon really does not directly concern its origin. However, some aspects of the story of the Moon's evolution do merge into the story of its origin. Where did the heat come from to make the magma ocean? Was there a cataclysm, and if so, did it have something to do with accretion? The Moon is locked into synchronous rotation with the Earth (i.e., its "day" of 709 hours approximately equals its "year," the time it takes to revolve once around the Earth, see Chapter 1). Is this fact, as well as the Moon's curious orbital properties, related to its origin? These questions alone are enough to make the origin of the Moon of interest to lunar geologists.

Many people think that we have "solved" the problem of lunar origin. It is claimed that the "giant impact" model (a.k.a., the "Big Whack" or, in a more dignified mode, the "Collisional Ejection Hypothesis") has made earlier models of origin obsolete. All that remains to be done is to mop up the mere details of explaining the facts about the Moon and its history. Perhaps this is a solved problem, but I hope to show in this chapter some of the reasons the Big Whack model has such wide appeal and to suggest that perhaps we are not quite so smart as we tend to think we are.

The Moon Cannot Exist!: Early Models for Lunar Origin

Ignoring for the moment the mythological stories of the birth of the Moon, lunar origin was first considered as a scientific problem by the classical theories of the origin of the solar system in general. The dominant model for the origin of the solar system holds that it condensed out of a cloud of hot gas, called the *solar nebula*. In such a model the planets are small pieces of the condensed material left over after star formation. Thus our Moon and the moons of the giant planets are similar to the planets that form around a star, a moon being the leftover bits of planet formation. This model, now called the *binary accretion model* (or the co-accretion model), was strongly defended by the astronomer Edouard Roche (who developed the idea of the tidal breakup of large bodies). In the co-accretion model of origin, the Moon is sometimes referred to as Earth's "sibling" because they both accumulated and grew as separate planetary objects in orbit around the Sun (Fig. 7.1).

As mentioned earlier, the Moon is receding, or moving away from Earth over time. This recession is extremely slow, about 4 cm per year. The converse of a lunar recession is that the Moon must have been much closer to Earth in the geological past. The astronomer Pierre Simon Laplace, inventor of the solar nebula model, attempted to account for the Moon's recession from Earth and concluded that tidal interactions between Earth, the Moon, and the Sun would conspire to cause the Moon's retreat from Earth over time. A corollary of this idea is that while the Moon recedes, Earth's rate of rotation is decreasing. Thus the evolution of the orbital relations in the Earth-Moon system is an important constraint on models of origin.

One of the first scientific models for the origin of the Moon was developed by George Darwin, the son of the great biologist and geologist Charles Darwin. Darwin, following up on Laplace's suggestion that the Moon is gradually receding from Earth, traced this recession back in time and concluded that Earth and the Moon were originally in physical contact with each other. After this idea is combined with the concept that

Figure 7.1 (opposite). Contending models for the origin of the Moon. The three classical origin models are (a) the "spouse" (the capture model), (b) the "daughter" (the fission model), and (c) the "sibling" (the binary accretion, or co-accretion, model). They have now been superseded (or joined) by the "Big Whack" model, or the collisional ejection hypothesis.

proto-Moon's orbit around the Sun

Earth's orbit around the Sun

grazing encounter of proto-Moon with Earth

INTACT CAPTURE MODEL

a

Moon's orbit

proto-Earth's orbit around the Sun

Rapidly spinning molten proto-Earth

FISSION MODEL

b

Moon's orbit

proto-Earth's orbit around the Sun

debris in Earth orbit

Moon's orbit

CO-ACCRETION MODEL

c

Earth's rate of spin was once much greater, it is a short leap to the suggestion that the Moon originated by being flung off Earth very early in Earth's history, when it was spinning very rapidly (Fig. 7.1). We now refer to this concept as the *fission model* for lunar origin. Because the Moon came from Earth, the Moon is considered to be Earth's "offspring." An early version of this idea held that when the early Moon spun off from Earth, the scar left on Earth was the Pacific Ocean basin. We now know that the ocean basins are very young features (younger than 70 million years), but this knowledge, unavailable to 19th-century astronomers, does not negate the fission model—other facts do that.

The third classical model for lunar origin grew out of the recognition that some satellites, particularly a few of the moons of the outer planets, orbit their primaries in a retrograde (opposite to planet rotation) direction. These orbits indicate that the small moons are wanderers in space that have been captured by their planets. The American astronomer Thomas Jefferson Jackson See, notable for claiming to be able to see craters on the surface of Mercury through the telescope (they are there but are too small to be visible from Earth), advocated that our Moon was captured into Earth orbit early in Earth's history. See claimed that the Moon had formed in the outer solar system and had wandered closer to the Sun, as had all of the planets through time. Eventually, some perturbation, probably an encounter with Jupiter, had flung the Moon into the inner solar system, where it was captured by Earth. The *capture model* (Fig. 7.1), modified appropriately to account for modern ideas about varied compositions in different parts of the solar system, was the third model of lunar origin to be tested by the data from the Apollo missions. In the capture model, the Moon is Earth's "spouse" (a forced marriage!).

What was Apollo's contribution to our understanding of the origin of the Moon? Essentially, the lunar samples and the Apollo program provided a variety of indisputable facts about the Moon's composition, the ages of its materials, and the reconstruction of certain past events—facts that any model for origin must explain. When these new facts are combined with certain astronomical data, we have information that can constrain the origin models. If a model for lunar origin makes predictions contrary to some known property of the Moon, it cannot be correct. If a model cannot explain a certain observation or if it accounts for many but not all of the observations, it may or may not be correct; most certainly it is incomplete. We should look at the properties of the Moon that any successful model of origin must explain and the constraints that such a model must satisfy.

First, let's look at some general facts about the Moon and its orbit. The Earth-Moon system as a whole possesses a great amount of *angular momentum*. This property of rotating systems includes not only the rates of spin but how much mass is spinning and how difficult it would be to stop such spinning. Angular momentum can be neither created nor destroyed—it can only be transferred. The Earth-Moon system has the greatest amount of angular momentum of all the planet-satellite systems in the solar system. The Moon's orbit is neither in the plane of the ecliptic (the plane in which nearly all the planets orbit the Sun) nor in the equatorial plane of Earth (Fig. 1.6). The spin axes of Earth and the Moon are not aligned. Earth's axis is inclined 23.5° from the ecliptic plane (the plane of its orbit around the Sun), whereas the Moon's axis is nearly vertical, being inclined only 1.5° from the ecliptic plane. These physical properties of the Earth-Moon system do not allow the dynamic requirements of some origin models, such as the capture model, to be easily satisfied.

Another constraint on lunar origin is the bulk composition of the Moon. Estimating the composition of an entire planet is tricky. Because we know approximately what materials the planets originally assembled from (asteroids similar in composition to primitive meteorites), the density of a planet tells us something about its bulk composition. The Moon's density is about 3.3 g/cm³ (grams per cubic centimeter), much less than Earth's density, at 5.5 g/cm³ (but about the same as the density of the *mantle* of Earth). The most straightforward explanation for this difference is to assume that the Moon *as a whole* is depleted in iron, the most abundant high-density element in the solar system. Note that this bulk depletion in iron has nothing to do with the iron-rich nature of the mare basalts or certain other lunar rocks. The density data mean that the *whole* Moon has less iron than Earth. We think that this is because Earth has a massive core of iron-nickel metal, whereas the Moon has either no core or a very small one. Estimating the amounts of the other elements depends on knowing how and when the Moon melted and how these magma compositions changed during cooling. For that information, we needed data from the Apollo missions.

The legacy of Apollo has added significant details to our constraints on lunar origin. We have already noted that lunar rocks have no detectable water and very small amounts of elements that we consider to be "volatile" (i.e., have low boiling temperatures). The volatile hydrogen and helium that we find in the lunar soil come from the Sun. Note that the lunar samples are *depleted* in volatile elements (relative to Earth), not *devoid* of them. The distinction is important. Whatever process was responsible for creating the Moon, it could not totally eliminate all of the volatile elements. The bulk Moon is probably enriched in more *refractory* elements (those with high boiling points) than typical planetary material. The age of the Moon, as inferred from its content of radiogenic elements and isotopes, is about the same as the age of Earth, 4.5 billion years old.

Most elements have more than one stable isotope. For the Moon, study of the nonradioactive, stable isotopes of oxygen show that the Moon and Earth are closely related (Fig. 7.2). Because meteorites that have formed in different parts of the original solar nebula have different oxygen isotope compositions, the data from samples imply that Earth and the Moon must have formed in about the same place in the solar nebula, at roughly the same distance from the Sun. Additionally, the evidence from the samples indicates that the Moon had an ocean of magma early in its history. Such an episode of global melting requires a heat source, indicating that the initial formation of the Moon must have been a high-energy process.

None of the three classical models of lunar origin are particularly successful at satisfying these constraints or at explaining the properties of the Moon. In fact it has been said that because none of the models of origin are outstandingly successful, perhaps the Moon cannot exist! Although the fission model is supported by the depletion of the Moon in iron and volatile elements and by the oxygen isotope relations of Earth and the Moon, it has difficulty explaining the dynamics of the Moon's

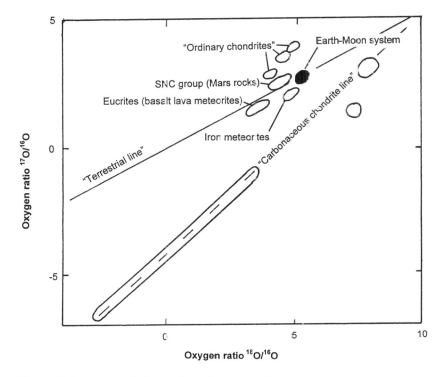

Figure 7.2. A map of the various isotopes of oxygen, a very common element in rocks, showing where various materials in the solar system formed. Whereas all known meteorite samples group into clusters, indicating that they formed at various locations around the solar system, the Earth-Moon samples make up a single cluster, indicating that however the Moon formed, it is closely related to Earth and was created in the same part of the solar system.

orbit and its orbital evolution. The capture model does not adequately account for these same dynamics and additionally has difficulty explaining the similarities in the oxygen isotopes (why are they the same if the Moon came from elsewhere in the solar system?). The co-accretion model perhaps does the best job of accounting for lunar properties, but even it does not really explain the Moon's depletion in volatile elements and in fact may be *too* flexible (a model that can be stretched to fit any data does not really explain anything). Even so, the co-accretion model was the favorite model of origin for most scientists in the years immediately after Apollo, probably because it was the least objectionable one of an admittedly poor lot.

This was the state of affairs when some lunar scientists decided to convene a conference on the origin of the Moon so that quiet contemplation of this thorny problem in the midst of active intellectual stimulation might produce some new insight into the problem of lunar origin. The conference was held in 1984 in Kona, Hawaii, and amazingly enough, it produced the hoped-for synergy (and possible breakthrough) in thought.

Giant Impacts and Rare Events: The Big Whack

The idea that very large impacts occurred early in the history of the planets grew out of two separate lines of thought. Scientists making theoretical models of the dynamics of planetary accretion found that nothing prevented the formation of fairly large planetoids (hundreds of kilometers across) or the subsequent collision of these objects with and incorporation into the existing planets. Some of these early planetoids could have approached sizes comparable to those of the surviving planets. Other scientists mapping the geology of the Moon and planets noted that the largest craters on these bodies were very large indeed; the South Pole–Aitken basin on the Moon is over 2,500 km across (about the distance between Houston and Los Angeles), and another supposed feature, the now discredited Procellarum basin, was alleged to be over 3,200 km in diameter (about the distance between New York and Phoenix). It required no great leap of faith to connect these two disparate observations into the concept that early planetary accretion involved very large impacts or, perhaps a better term, "planetary collisions." Such events might or might not have actually happened. Like the formation of large basins, a planetary collision was literally "hit or miss."

The idea that the Moon formed as a consequence of a giant impact on Earth was originally proposed independently by two research groups in 1976. That idea lay fallow until the Kona conference, when it exploded into the consciousness of the community. In one fell swoop the giant impact model seemed to have solved many problems of lunar origin. An off-center impact could have simultaneously increased the spin rate of Earth and launched material into Earth orbit. An oblique collision also would have greatly increased the angular momentum of Earth and its newly added partner, accounting for one of the most curious of the dynamical properties of the Earth-Moon system. The ejected matter from the collision would have been superheated vapor, material that would certainly be depleted in, if not devoid of, volatile elements. The Moon's lack of iron could be the result of the fact that the colliding planet had already separated into core and mantle. Thus the material out of which the Moon formed came from the mantles of the two bodies—the proto-Earth and the colliding planetoid. Both of these parts of the planets would already have been depleted in iron, having removed most of their iron to their respective cores.

The Big Whack explains the Moon as debris ejected from a planetary collision. About 4.5 billion years ago, two planets were in the position around the Sun now occupied by the Earth-Moon system. The "proto-Earth" (Terra before the impact) was already planet-sized (only slightly smaller than the present Earth, 12,756 km in diameter), had differentiated into a core and mantle, and rotated at a much slower rate than Earth does currently. The other planet (the "impactor") was also differentiated into core and mantle and was about the size of Mars (6,787 km in diameter). The two planets collided off-center (Fig. 7.3); the impact strike by the Mars-size body was in the same direction as the rotation of the proto-Earth, resulting in a speeding-up of Earth's rate of rotation. But more important, a vaporized cloud of debris was flung off the colliding planets and into orbit around Earth (Fig. 7.3). It is still uncertain how much of this material came from each object, but current models suggest that most of the ejected material came from the impactor, the currently nonexistent, Mars-sized planet.

This superheated cloud of vapor expanded rapidly and cooled in space, forming an orbiting cloud of small, recondensed droplets of vaporized mantle material. One might expect such droplets to be enriched in refractory elements (those with very high boiling temperatures), having formed at extremely high temperatures and being depleted in volatile elements. The Moon is made from this material. A cloud of debris orbiting Earth would

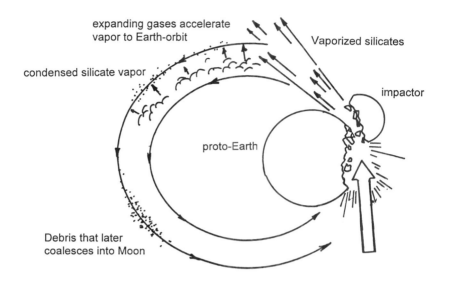

expanding gases accelerate
vapor to Earth-orbit

Vaporized silicates

condensed silicate vapor

impactor

proto-Earth

Debris that later
coalesces into Moon

GIANT IMPACT MODEL

Figure 7.3. Diagram showing how the Big Whack model launches vaporized debris into orbit around Earth. An off-center, oblique impact seems to be required for this model. This vaporized rock later cools into many very tiny droplets that later accumulate very rapidly into a single body. According to advocates of this model, the rapid accumulation is responsible for the heat that created the magma ocean. Solid material would follow the inner ballistic path, reaccreting onto Earth; this material includes most of the impactor. After J. Wood, "Moon over Mauna Loa: A Review of Hypotheses of Formation of Earth's Moon," in Hartmann, Phillips, and Taylor, *Origin of the Moon,* 17–55

not be stable as an extended body, and this debris would have assembled itself into the Moon on very short timescales, tens of thousands of years, according to the dynamical models. Accretion within such short periods released large amounts of thermal energy, and it was this energy that was responsible for the creation of the magma ocean. The bulk of the mass of the impacting planet was incorporated into Earth. Such an incorporation might incidentally explain a curious (and until now apparently unrelated) fact about Earth's upper mantle: it appears to have rather high concentrations of the siderophile ("iron-loving") elements, such as nickel, cobalt, and iridium. In the Big Whack model this elevated content of siderophiles is the result of the incorporation of the core of the impactor planet into the mantle of Earth.

By simultaneously explaining so many properties of the Moon and its motions, this model of origin became an instant hit. Numerical modelers rushed to run new, ever more powerful mathematical simulations of giant impacts on their new supercomputers. Geochemists assembled data on the abundance of ever more obscure elements, like tungsten and molybdenum. Geophysicists reexamined their meager data on the interior to compute new and better models for the bulk composition of the Moon. Geologists assembled new geological maps showing that giant basins and, by implication, giant impacts were even more common on the planets than previously thought. All were eager to jump on the Big Whack bandwagon.

The Big Whack not only appears to solve many problems about the origin of the Moon but also appears to solve other problems of planetary science. The planet Mercury appears to be exceedingly dense, about 5.4 g/cm^3, suggesting that it has a relatively large amount of iron. Reasonable models for the structure of Mercury indicate a very large core, almost 60 percent of the volume of the planet. Venus rotates on its axis very slowly, once every 266 days, and backwards. Put these facts together with the Big Whack model and voilá—problem solved! Mercury and Venus are fragments of a giant impact, but in the opposite sense of the Earth-Moon collision. This impact reversed and slowed the rotation rate of Venus, stripped off much of the silicate mantle of Mercury (leaving the residuum of a planet, accounting for its anomalously large iron core), and sent Mercury careening in toward the Sun to become the innermost planet of the solar system. In another example, the spin axis of Uranus is tilted over 90° from perpendicular to the ecliptic plane. This extreme amount of tilt is proposed as a likely result of a giant impact early in the history of Uranus.

The Big Whack idea is now accepted wisdom in planetary science. It is seldom that a model gains such widespread acceptance so rapidly. It has done so largely because it readily explains so many different aspects of the Moon and its history.

Diverse phenomena from geochemistry to orbital dynamics appear to be accounted for. But is this because they really have been explained, or have they been explained away?

A Solved Problem?

Despite the popularity of the Big Whack model, it is far from certain that we have solved the problem of lunar origin. In science no model is ever "proven" to be true. Instead ideas (hypotheses) are created and then tested repeatedly. If an idea passes these tests, it becomes generally accepted. Any good hypothesis not only will explain existing facts but also will make predictions that are themselves testable and incidentally explain apparently unrelated facts. For a hypothesis to be a valid one, it must be capable of being "falsified," that is, while a hypothesis cannot be proven to be true, it must be capable of being proven untrue. We always have more error than truth in the multitude of hypotheses and models that scientific research produces, and unless this criteria is adhered to, we have no way to discard unusable, wrong concepts in favor of more correct ones.

One of the biggest difficulties with the Big Whack model is its elasticity. It has so many loose ends that it can be stretched to fit many observed facts. A model so unconstrained loses its predictive potency. For example, in terms of geochemistry, the Big Whack model (as outlined above) is actually a variant of the capture hypothesis, a variant that solves the dynamical problems that made the conventional capture unworkable. But it is not clear how geochemical data can constrain the model. The material that makes up the Moon comes from elsewhere in the solar system. If the Mars-sized impactor came from somewhere else, why do the oxygen isotopes of Earth and the Moon match so closely? Advocates of the Big Whack explain this problem away by saying that the isotopic data cannot be overinterpreted and that isotopic affinities merely indicate where materials are created in the solar system in general. If so, then perhaps the oxygen isotope data are not really constraints on origin at all.

Another difficulty with the Big Whack is the nature of the Moon's depletion in volatile elements. It is generally thought that this depletion is one of the lunar facets most readily explained by the Big Whack. However, the patterns of elemental depletion do not seem to make sense. For example, some volatile elements clearly were incorporated into the accreting Moon, as evidenced by the gases driving the eruptions of dark mantle material (ash) and the vesicles in the mare basalt lavas. How were these gases preserved in such a high-temperature environment? Some elements (such as manganese) are found in the samples that are volatile in a geochemical sense, yet these elements are not as depleted as other volatile elements of comparable geochemical behavior. Is this also a consequence of the Big Whack? Clearly, the effects of volatile depletion are not understood well enough to use them as a real constraint on the processes operating during giant impacts.

My discussion of these difficulties does not aim to disparage the Big Whack model but to suggest that we do not yet fully comprehend the origin of the Moon. In particular we need to understand how such an event would affect properties that we can determine and measure to a precision adequate to allow us to really test the model. If we cannot devise rigorous tests of such a hypothesis, then it becomes merely an interesting idea, not a generalized theory of lunar origin. The Big Whack model is the most promising idea yet developed to explain some of the more puzzling aspects of the origin of the Moon. Future work on all of its various components will enable us to better assess its true value.

 Chapter 8

A Return to the Moon

When the *Apollo 17* spacecraft departed in December 1972, the Moon returned to the quiet solitude it had enjoyed for most of its history. Until recently, except for brief visits by Soviet robotic probes (including the *Luna 24* sample-return mission in August 1976), no one had been back to the Moon, and it appeared that no one would be back for the foreseeable future. Ever since we left the Moon, there has been talk of returning. For most of the last 20 years, this discussion has focused on the Lunar Polar Orbiter (LPO) mission, which planned a robotic spacecraft that would measure the chemistry, mineralogy, gravity, and topography of the Moon. The LPO mission would help us to follow up on the discoveries made by the Apollo missions. The LPO was never started, despite a halfhearted request by NASA during its lackluster campaign for the now-defunct Space Exploration Initiative.

Why have we stopped going to the Moon? Is it because there is nothing significant there to see or nothing important to do? Is it because of a lack of resources, the needed funds being unavailable? What are the prospects for a return to the Moon? What would we do if we could go back? In this chapter I want to look at the numerous efforts, made by a number of vocal advocates, to return to the Moon and why such efforts repeatedly failed. We will then examine our one success in the field of lunar return, the Clementine mission, flown in early 1994, and discuss why it succeeded where others had failed.

The Whole Moon Catalog: The Lunar Polar Orbiter Mission

The data returned by the orbiting Command-Service Modules on the last three missions to the Moon (*Apollos 15, 16,* and *17*)

were tantalizing hints of what could be accomplished scientif-
cally from lunar orbit. On these missions we measured the chemistry of the surface, sniffed for gas emissions from the Moon,
tracked the spacecraft to measure the gravity field, saw strange
magnetic anomalies from orbit, determined the topography of
the surface by laser ranging, and took photographs of unprecedented clarity and quality, showing the surface in greater detail
than ever before. The data from these advanced sensors were
superb—the problem was that they covered only a small part of
the Moon, near the equator. For purposes of safety and accessibil-
ity, the Apollo missions were confined to orbits close to the lunar
equator. Thus the higher latitudes remained largely unknown
territory.

The solution to this problem was understood: Put a spacecraft
into *polar* orbit around the Moon. Because the plane of a spacecraft's orbit remains fixed in space, as the Moon slowly rotates
on its axis every 29.5 days the groundtrack (the trace of an orbiting spacecraft on the surface of a planet) will cover the entire
globe in the course of one lunar day. In such a mission profile we
could map 100 percent of the surface in a mission lasting only a
single month. This was not a new insight. The last two Lunar
Orbiter spacecraft (*Orbiters 4* and *5*) were put into near-polar
orbit for just this purpose—to photograph sites of scientific interest far from the equatorial latitudes, a mission objective that
was brilliantly accomplished (e.g., Schrödinger basin at 76° S
latitude; Fig. 2.7). In the mission scheme for Lunar Polar Orbiter
(Fig. 8.1), we would not only photograph the surface globally but
also measure the surface composition and gravity to provide a
regional and global context for understanding what the Moon
rocks were telling us.

How is it possible to measure the composition of the surface
from orbit? A variety of techniques remotely measure composition. Each technique tells something different, and distinct parts
of the electromagnetic spectrum contain different, useful geological information (Fig. 8.2). The Moon has no atmosphere or global
magnetic field; thus all wavelengths of radiation strike its surface, including cosmic rays from both the Sun and the galaxy and
X-rays from the Sun. When this radiation hits the Moon, it induces energy to be reradiated from the Moon's surface. Such
reradiation from the lunar surface has energies and wavelengths

Figure 8.1. The Lunar Polar Orbiter spacecraft in one of its many incarnations, *Lunar Observer,* a proposed mission to globally map the composition and surface of the Moon with remote-sensing instruments. This mission was studied extensively, yet it never flew.

characteristic of the types and amounts of the elements in the surface. It is this principle that permits us to determine the chemical composition of the Moon from orbit.

The LPO spacecraft would have carried two instruments to measure this radiation from the surface: X-ray and gamma-ray spectrometers (Fig. 8.2). The X-ray spectrometer uses the Sun as an X-ray source; thus it can determine composition only on the daylight (Sun-lit) side of the Moon at any given time. The X-ray spectrometer measures the concentration of elements with low atomic numbers, including aluminum, silicon, and magnesium, three important rock-forming elements on the Moon. The concentrations of heavier elements are measured by a gamma-ray spectrometer, which measures extremely short wavelength radiation (Fig. 8.2). Gamma radiation from the lunar surface

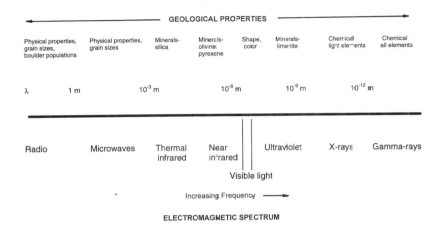

Figure 8.2. A diagram of the electromagnetic spectrum, showing the different geological information that is obtained from each portion of the spectrum. Basically, the information contained at various wavelengths is what permits the determination of composition and physical properties, the field of "remote sensing."

comes from two sources. The first source is natural radioactivity, in which certain elements decay into other elements and emit gamma-rays. An example is uranium, a radioactive element naturally present in the Moon (at extremely low levels). Thus in this mode, the instrument is basically a big Geiger counter in space. The second source of gamma-rays is caused by the bombardment of the surface by cosmic rays. This bombardment induces reradiation through the interactions of the gamma-rays with the nuclei of lunar atoms. Thus we can measure the concentration of elements such as iron, titanium, and the rare earth elements by carefully examining the *spectra* (energy intensity as a function of wavelength) of the reemitted gamma radiation from the Moon. Together, the X-ray and gamma-ray instruments measure almost the complete major- and minor-element composition of the surface from orbit.

It is important not only to know the concentration of the elements in lunar materials but also to understand how those elements are arranged with respect to each other, in other words, the *mineralogy* of the Moon. It turns out that by very precisely measuring the color of the surface, we can determine its mineralogy. Light from the Sun interacts with a mineral in a way that is characteristic of the crystal structure of the mineral. At certain wavelengths of light (which *is* color), the crystals absorb light energy, resulting in a suppression of the amount of light reflected. This effect is called an *absorption* and results in a "dip" or "band" in the reflectance curve (Fig. 8.3). *Absorption bands* have unique shapes, depending on the mineral, and a complex mixture of minerals (a rock) will likewise have a mixed absorption spectrum. By studying the returned Apollo samples, we know what different Moon rocks will look like at different colors. Thus with an instrument that maps the color from orbit, we can map the distribution of rock types on the Moon.

Lunar color is very subtle (differences are typically less than a few percent of reflectance), so the instrument must be capable of measuring these very tiny differences. This technique, called *reflectance spectroscopy*, is applied to the Moon from Earth-based telescopes (in fact this is how we learned the method for mapping minerals remotely), and from these studies, we have learned many fascinating facts about the way rocks are distributed on the Moon. However, from Earth we cannot see the far side, so our coverage is not global, and we are limited in the size of features seen, the smallest being about 1–2 km in size at best. Making these maps from orbit is important to understanding the whole Moon at scales appropriate for geological mapping.

In addition to mapping rocks, the LPO mission would have completed the photographic mapping of the Moon, taking stereo images so that we could make topographic maps of the surface. We need to know topography because the shape and the size of features are clues to the surface processes responsible for their creation. The topographic maps made by stereo photography need to be tied together regionally to provide global data. An easy way to do this is to make precise range measurements from the orbiting spacecraft. When the orbit is subtracted from the range data, we have a profile of the topography of the surface. Such data show the global *figure* or shape of the entire Moon. The figure of the Moon is an important clue for a variety of internal processes, including its thermal history.

Another set of clues to lunar history is provided by detailed gravity maps of the surface (Fig. 8.4). You will recall that the

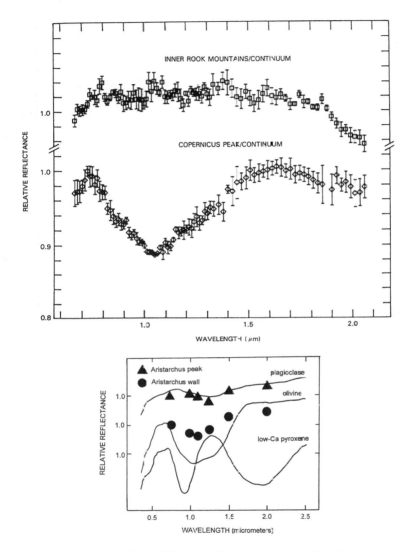

Figure 8.3. Top: examples of lunar reflectance spectra for the
mountains of the Orientale basin and for the central peak of the
crater Copernicus on the Moon, as seen from telescopes on Earth.
Bottom: laboratory spectra for common, rock-forming minerals
(lines) and for the peak (triangles) and wall (circles) of the crater
Aristarchus, as seen from the orbiting *Clementine* spacecraft. These
data show that the peak of Aristarchus is nearly pure plagioclase (the
rock type anorthosite—see Plate 11), whereas the wall is made up of
an olivine-plagioclase mixture (called *troctolite*—see Plate 12). Thus,
from the spectra of known samples, we can map the mineral
composition of unvisited sites on the Moon.

Figure 8.4. The gravity field of the Moon, mapped at low resolution by the *Clementine* spacecraft. Large increases in gravitational attraction are associated with the circular basins. These features are the mascons (mass concentrations) and indicate areas where the dense, massive mantle has been uplifted close to the surface by basin-forming impacts. Courtesy of Maria Zuber, Johns Hopkins University.

discovery of the mascons from mapping the gravity field through the tracking data from Lunar Orbiter spacecraft was surprising (see Chapter 3) and led to a revision of models of the internal density structure of the Moon. Unfortunately, we have this information for only one-half the Moon because the Orbiter spacecraft were out of sight when flying over the far side. How can we track the spacecraft when it is out of sight? The answer is to use another spacecraft to relay the radio signals to Earth. This relay satellite (called a *subsatellite* because it would be launched from the LPO spacecraft in lunar orbit) would be in a very high orbit where it could almost constantly see Earth and would have only one function: to relay to Earth the radio signal from the LPO while the spacecraft is behind the Moon. From these data, we could produce a detailed gravity map for the entire surface.

Various other instruments were proposed to measure the magnetic fields and heat flow of the Moon and to characterize the lunar "atmosphere." Although the Moon has no global magnetic field, its surface is dotted with small zones where the crust is strongly magnetized. These magnetic zones may be caused by unusual impact conditions or subsurface geology (see Chapter 2). Mapping the anomalies from orbit would allow us to determine the likely causes of lunar magnetism by seeing which features the anomalies are associated with. Heat flow tells us about the bulk composition of the Moon. It is important to map the variations in heat flow globally to determine the abundance of heat-producing elements in the Moon. We can determine heat flow by measuring the surface temperature at long (microwave) wavelengths from orbit. The Moon has no atmosphere, but it is surrounded by an extremely tenuous cloud of vapor, made up of exotic materials such as gaseous sodium. This "atmosphere" has an extremely low density (the rocket exhaust from each Apollo landing temporarily *doubled* the global mass of the lunar atmosphere). The lunar atmosphere is imperfectly understood. Instruments in orbit could measure the density and composition of these gases and their variation as a function of position and time.

Many scientific committees and study groups, both internal to NASA and external from it (such as the Committee for Lunar and Planetary Exploration of the National Academy of Sciences), formed and met over a period of 20 years, each one recommending that the LPO mission be flown and be flown soon. Yet from 1973 to 1989, not a single NASA budget request contained the LPO mission as a new start. The mission was studied and restudied. It was broken up into several smaller pieces and then recombined into one mission. It was downsized, "descoped," and rescaled. It was justified, defended, and rejustified. Everything possible happened to it except for one thing: It was not flown.

Lunar Bases and Space Nonactivities of the 21st Century

After the national goal of the lunar landing was accomplished, the country had no long-term goal for its space program. A commission established by the White House in 1969 recommended that we undertake a long-term program to conduct a human mission to Mars by the mid-1980s. For a nation mired down in the Vietnam War and strongly influenced by antitechnology social engineers, such a proposition had no chance of implementation. Instead NASA settled for the construction of a space transportation system (the Space Shuttle), designed to make access to low Earth orbit routine and to decrease the "high" cost of spaceflight typified by the "disposable" *Saturn 5* booster rocket that took Apollo spacecraft to the Moon but was soon rendered obsolete by the supposedly more advanced, cost-effective Shuttle.

As it turned out, the Shuttle took longer to develop than planned and became gradually less capable as NASA was pushed to lower the initial buy-in costs at the expense of stretching the program over a longer time (the same logic is now applied to the Space Station). The original concept for the Space Shuttle as envisioned by Wernher von Braun, the architect of much of the nation's space capability, was as the first piece of a space-faring infrastructure, including a station in low Earth orbit, transfer vehicles, lunar ferries, and ultimately, a manned interplanetary vehicle. Von Braun visualized a completely reusable Shuttle, employing liquid-fuel rockets throughout, including a piloted booster stage. However, to save money, NASA designed and built the Shuttle with two strap-on solid-fuel motors, a cumbersome and dangerous arrangement (once started, solid-fuel rockets cannot be shut down).

The Space Shuttle program was approved by a grudging Congress and was slowly built by a struggling NASA during the remainder of the 1970s. It finally took off in April 1981, ably flown by veteran astronaut and *Apollo 16* "moonwalker" John Young. With the Shuttle flying, NASA had to sell the next logical step in space to a skeptical Congress. President Ronald Reagan was convinced of the value of the space program, as much for its inspirational power to the American people as for anything else. With Kennedyesque echoes, Reagan announced in his January 1984 State of the Union Address that he was directing NASA to build a permanent manned space station, the second piece in Von Braun's vision of a space-faring infrastructure, and to "do it in a decade." Although not widely embraced by everyone in the space program (especially by an endlessly complaining community of space scientists who saw their own programs threatened by this new effort), the Space Station became the central core of NASA's program, the rationale for the long-term existence of the agency.

President Reagan's deadline came and went with the Space Station nowhere in sight. What is significant to our story is what happened in the early 1980s. One of the components planned for the Space Station after its initial construction was the Orbital Transfer Vehicle (OTV). The OTV was designed to ferry people and equipment between the Space Station in low Earth orbit and a spot in what is called *geostationary orbit*, a perfectly equatorial orbit 22,000 miles above Earth. From such a position, satellites orbit once every 24 hours, the same time it takes Earth to rotate, and they thus appear to "hover" over a single spot on the ground. These orbits are highly desirable spots for weather and communications satellites.

At the Johnson Space Center in Houston, veteran Apollo scientists Michael Duke and Wendell Mendell had an epiphany. Because it takes almost the same amount of transfer energy to go from the Space Station (in low Earth orbit) to geostationary orbit as it does to go from the Space Station to lunar orbit, they realized that once the OTV was in operation, we would have the ability to go back to the Moon. If the ability existed, it would probably be used. Duke and Mendell began to think about the consequences of a return to the Moon, bringing engineers and scientists from NASA, national laboratories, and universities into a "lunar underground," an informal network of people dedicated to returning to the Moon and establishing a lunar base.

After a workshop in Los Alamos, over 200 enthusiastic people met at the first Lunar Base Symposium in Washington, D.C., at the National Academy of Sciences in April 1984. There was little attempt at this meeting to "justify" a return to the Moon. It was merely assumed that one day, we *would* return, and given that assumption, what would we do when this happened? From this conference and the Los Alamos workshop was born a "Lunar Base Initiative," which successfully lobbied the presidential National Commission on Space to include, in its final report in 1986, a lunar base as a future national activity in space. In response NASA created a new office, the Office of Exploration, designed specifically to examine the ways in which we could conduct both a return to the Moon and the first human missions to Mars.

Over the next few years mountains of paper were generated, devising strategies, scenarios, and imaginary missions. Much of this work found its way onto overhead transparencies (viewgraphs), NASA's preferred mode of communicating ideas internally. A favorite gag of insiders suggested that we could reach the Moon merely by stacking all of our viewgraphs on top of each other! A typical activity was the design of an *architecture,* jargon meaning the way a mission goal is approached. For example, the mission profile of the lunar-orbit rendezvous used by Apollo is an example of an architecture. Architectures include not only mission plans and rockets but also the supporting elements (such as launch and tracking facilities) needed to conduct a mission. Devising architectures and producing paper studies formed a relatively "safe" way to blow off lunar base steam, in the absence of any long-term national space goal.

The tragic *Challenger* Space Shuttle explosion in January 1986 stopped the space program in its tracks. An extensive investigation was conducted to fully understand the accident and its causes. NASA's competence was seriously called into question for the first time since the tragic *Apollo 1* fire of 1967. It was more than two years before the Shuttle flew again, the longest single stretch of time between a space accident and flight resumption since the space age had begun. Additionally, the *Challenger* accident froze other space plans being studied at the time, a chill not totally thrown off after the Shuttle began to fly again.

Early in the mission studies of the late 1980s it became apparent that at least one robotic precursor, namely the LPO, was a desirable mission to increase our general strategic knowledge about the Moon before we returned there. If nothing else, having global data would allow us to be sure that we placed the site of the lunar base at the best-possible location. Another development concurrent with these studies was a surge of interest in using the resources of the Moon. The object of this research was to understand how we might be able to "live off the land" when we returned to the Moon. As a result, many of the architectures for lunar return began to incorporate the use of local resources, sometimes at unrealistic levels.

Circumstances seemed to change significantly on July 20, 1989, the 20th anniversary of the *Apollo 11* landing. In a speech at the National Air and Space Museum, President George Bush challenged the nation with a new vision and a new and much-needed focus for its space effort: the completion of the Space Station; a return to the Moon, this time to stay; and a manned mission to Mars. The plan followed the sequence outlined by Wernher von Braun 40 years earlier: shuttle, space station, lunar ferry, Mars mission. To Von Braun, Apollo had been an anomaly, in which the country had been forced to jump out of the logical sequence in space because of political necessity. Once the political objectives had been satisfied, he thought that it was logical to return to the master plan.

The Space Exploration Initiative (SEI)—a.k.a., the Lunar-Mars Initiative, the Human Exploration Initiative, the Moon-Mars Initiative—began an intensive round of study. NASA responded to the presidential call with its own internal report, referred to as the "90-Day Study." That effort outlined five different approaches (called "options") for a return to the Moon and for a human mission to Mars. Each architecture used Space Station *Freedom* as a transport station in Earth orbit, used Shuttle- and Station-derived hardware and modules, and outlined a 30-year plan to accomplish these national goals. The internal agency estimate of the cost of these plans, an estimate not included in the report but leaked to the press and widely quoted, was between 500 and 600 *billion* dollars.

Needless to say, this report generated a lot of negative comments. Various pressure groups in Washington, which never had liked the idea of lunar bases or Mars missions, and congressional critics and their accomplices in the media began a campaign against SEI, trumpeting the $500 billion cost as their principal (but not only) complaint. Lowell Wood (a protégé of Edward Teller, father of the H-bomb) and his colleagues at Lawrence Livermore National Laboratory responded to the NASA "90-Day Study" with an architecture that called for the use of inflatable spacecraft (whose use had first been proposed by Von Braun in the 1940s) and an accelerated timetable that had us reaching

Mars within 10–15 years; it would cost only a few tens of billions of dollars.

The White House became concerned that SEI might be foundering because of NASA bungling, congressional indifference, and media hostility, and so in 1990 it set up two *ad hoc* study panels. One was devised to examine the nation's future in space and to recommend a general course of action (the Augustine Committee, named after its chairman, Martin-Marietta CEO Norman Augustine). The other group was established to evaluate the SEI mission architectures pouring into Washington from around the country. This panel, called the Synthesis Group, was chaired by *Apollo 10* astronaut Thomas Stafford and spent a year looking at various ways to implement SEI. The reports of these two groups endorsed the concept of SEI but differed on the way to do it: The Augustine Committee offered a "pay-as-you-go" philosophy (translation: we can't afford it), whereas the Synthesis Group suggested four architectures emphasizing different "themes," including human habitation, science, and resources (translation: why are we doing this?). The Synthesis report was released to a response of massive indifference in May 1991.

Congress effectively killed SEI during the budget process of 1991 and 1992 by "zeroing it out" (i.e., refusing to appropriate any funds for it). This action also wiped out the new start for Lunar Polar Orbiter; after 20 years of detailed study, the LPO mission was "dead on arrival" in its first appearance in a NASA budget request. These events were followed by the election defeat of President Bush, the originator and advocate of SEI. With no political support, a return to the Moon is farther away than ever. For now, building a base on the Moon is not being planned for or worked toward. NASA's main goal for the future (at the time of this writing) is to complete, sometime before 2002, the International Space Station. The reader might reflect on the events in our space program over the past 20 years and compare them with the events that occurred between 1961 and 1969. For the last two decades, we have spent much more money and accomplished far less in space than we did during those first eight years. There is no relief in sight. And the Moon awaits silently. . . .

While No One Was Looking: The Clementine Mission

The biggest and most innovative program of space technology development during the 1980s was not the civilian but the military space program. The nation has always had two parallel space efforts: From the beginning, the main impetus for the race to the Moon was national security. In the 1980s, research efforts in the military space program centered on the needs of space-based strategic defense, the Strategic Defense Initiative (SDI), or "Star Wars." A concept studied in SDI was the deployment of thousands of lightweight, high-technology sensors in a myriad of small satellites, a project referred to as "Brilliant Eyes," the brainchild of Edward Teller and his crew at Lawrence Livermore Laboratory.

One of SDI's development problems concerned systems tests because the 1970 Anti-Ballistic Missile Treaty prohibited tests of space-based defense hardware in low Earth orbit. During a discussion with colleagues in a Washington bar in 1989, Stewart Nozette of Livermore sketched out the concept for a mission that could test and qualify SDI technology in space and still avoid treaty problems: Send the spacecraft into *deep* space and use the Moon and an asteroid as targets for sensor tests. A quick study effort by an *ad hoc* team determined that the "Brilliant Eyes" sensors could indeed return useful scientific data for these bodies. Accordingly, NASA signed a memorandum of agreement with the Strategic Defense Initiative Organization (SDIO), Department of Defense, to conduct a joint mission to the Moon and to a near-Earth asteroid The goals of this mission were to space-qualify the SDIO hardware and to map the two objects with these small, advanced sensors. With a nod toward the old song "My Darling Clementine" (about the daughter of a miner in the 1849 California gold rush), the mission was christened "Clementine" because it would assess the mineral content of the Moon and of an asteroid, possibly with an eye toward "mining" those bodies in the future. After the asteroid flyby, the spacecraft would fly off into deep space, or just as the song says, "You are lost and gone forever."

In January 1992 I met veteran Apollo scientist Gene Shoemaker for dinner in a restaurant in Crystal City, Virginia, where he sketched out his concept of the Clementine mission. In our discussions, we realized that a superb scientific mission could be flown with this new technology, including not only global multispectral mapping but also laser altimetry to measure the topography of the Moon. The next day, we met with people from Lawrence Livermore Laboratory (a Department of Energy laboratory where the sensors were designed and built), SDIO, and NASA to review the mission concept and sensor set. Normally missions are studied for years before they gradually take shape and get started (for example, the LPO mission was studied for 20 years). Clementine was to be different: The object was to build, launch, and fly the complete mission within three years of its approval. The *Clementine* spacecraft was launched on January 25, 1994, *two years* after our meeting in Crystal City.

For many lunar scientists, this was a great opportunity, a chance to return to the Moon after years of nonaction. However, some in the planetary science community complained about the "militarization" of the space program and the "non-optimum" scientific nature of the instruments. Some were concerned that the mission had not been "studied" properly and that not enough scientists had been consulted to pronounce their blessings on the mission—in short, that no one had asked for their permission to fly this mission. Complaints were heard from some scientists that because *their* instruments weren't going to the Moon, it would be better that *no* mission be sent. As for myself and my colleagues, we were excited. We were actually going to get back to the Moon—and in our own lifetimes!

The *Clementine* spacecraft (Fig. 8.5) was built by the Naval Research Laboratory under the supervision of SDIO. Careful management by the superb program manager, Pedro Rustan, kept the spacecraft on schedule and on budget. Part of the reason the Clementine mission proceeded so well and so quickly is because it was a small project that was kept small; no more than about 300 people ever worked on the mission, even during times of maximum activity. In contrast, even the simplest NASA mission employs hundreds to thousands of people, many engaged in nonproductive activities. Clementine was handled the way NASA *used* to do space missions: fast, small, and on budget. The total cost of Clementine was about $80

Figure 8.5. The *Clementine* spacecraft. At 148 kg, this was a small but very capable spacecraft that mapped the Moon globally in 1994. The triangular area is the sensor deck, where the various cameras were mounted; a hinged door protected the cameras during maneuvers. The high-gain antenna is at the bottom, and the maneuvering rocket engine is at the top. Two solar panels provided electricity to the spacecraft.

million, including the launch vehicle, a surplus ICBM *Titan II* rocket. In contrast, the new NASA "faster, cheaper" mission series, Discovery, is capped at $150 million, *not* including the launch vehicle (about another $60 million).

After a month-long trip from Earth (on a circuitous route taken to conserve spacecraft fuel), *Clementine* arrived at the Moon on February 20, 1994, and spent the next two and one-half months orbiting and mapping the Moon. The mission was controlled from a renovated National Guard warehouse in Alexandria, Virginia, a facility named the "Batcave." A small team of Naval Research Laboratory engineers, led by Paul Regeon and the NASA-selected science team, including Gene Shoemaker and myself, worked around the clock, making sure that the data were being taken properly and marveling at our first global look at the Moon. After making 330 orbits and taking over 2.5 million pictures, *Clementine* blasted out of lunar orbit on May 3, 1994, on its way to an encounter with the asteroid Geographos, a meeting scheduled for August of that year. Unfortunately, a software malfunction in the spacecraft computer caused *Clementine* to spin out of control and forced the cancellation of the asteroid phase of the mission. The spinning spacecraft swung by the Moon on July 20, 1994, and entered solar orbit, 25 years to the day after *Apollo 11* landed on the Moon.

Clementine conducted our first global compositional and topographic mapping of the Moon. Some of the scientific results from the mission are discussed elsewhere in this book, in the appropriate topical sections. The global, digital, multispectral image of the Moon (Plate 14) was taken in 11 wavelengths, all filters being carefully selected by the science team to ensure that we could extract the appropriate mineral information. From these data we will be able to map the rock types that make up the crust. Because impact craters have excavated the Moon to a variety of depths, we can reconstruct the structure and composition of the crust in three dimensions. Strips of images taken by the high-resolution camera will allow us to interpret the surface process in many areas. We have infrared, thermal images of selected areas, allowing us to determine the physical properties of the surface layer at many different sites with diverse geological settings.

One of the most striking results from *Clementine* is the global topographic map we obtained from the laser altimetry (Plates 10, 15). For the first time, we can see the huge basins in all of their glory. The gigantic South Pole–Aitken basin on the far side was determined to be the biggest (2,500 km diameter), deepest (over 12 km) impact crater yet found in the solar system. Other degraded, almost obliterated impact basins stand out prominently in the topographic data. An astonishing result for some of these basins is their great depth; basins that appear nearly obliterated in the photographs seem to be as deep as they were the day they were created. This topographic information, combined with the compositional data provided by the multispectral maps (Plates 8, 9, 14), will allow us to probe the crust to great depths and will revolutionize our knowledge of the processes and history of the Moon.

Our first good view of the polar regions allowed us to identify areas of importance for future exploration. In particular, the *Clementine* view of the south polar region shows some interesting properties. There is a large area of darkness near the pole (Fig. 8.6), much greater in extent than would be predicted by the 1.5° tilt of the Moon's spin axis (Fig. 1.6). This dark region must be the result of the fact that the pole lies within the rim of the South Pole–Aitken basin, below the topographic level in which the Sun is visible. Dark regions near the pole never receive heat from sunlight and become "cold traps," zones in which the temperature does not exceed about −230°C, only 40° greater than absolute zero. If water molecules (for example, pieces of a comet) were to land in such cold traps, they could never get out again. Over geological time (the South Pole–Aitken basin dates back at least 4 billion years), significant amounts of water could have accumulated in the traps. *Clementine* did not carry instruments that allowed it to "see" into the dark areas, but an experiment improvised while *Clementine* was orbiting the Moon used the spacecraft transmitter to beam radio waves into the dark areas. Analysis of the reflected radio waves indicates that deposits of ice exist in the dark areas. This astonishing finding awaits confirmation by a future orbital or lander mission.

A small area near the south pole appears to protrude above the local horizon and is thus in almost constant sunlight. On the basis of *Clementine* data, this site is illuminated more than 80 percent of the lunar day during southern "winter" (the time of illumination would be even greater during the "summer"). The site, on the rim of a 20-km-diameter crater near the pole, would be an interesting one for a future landing. Such a site would permit the lander to use solar panels for electrical power. In addition, at about −30°C to −50°C, the site is thermally benign, lacking the temperature extremes experienced by sites on the equator: the heat of the lunar noon (over 100°C) and the cold of the lunar midnight (−120°C). The simple, nearly constant temperature here greatly simplifies the thermal design required of a spacecraft. Finally, the site's location among targets of such scientific interest (the South Pole–Aitken basin massifs, Fig. 6.6)

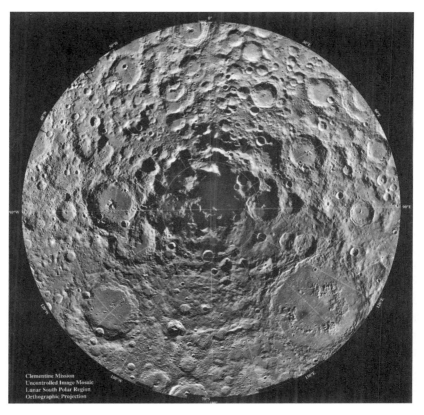

Figure 8.6. A mosaic of about 1,500 *Clementine* images of the south polar area of the Moon, from 70° S to the pole (center). The near side is the top half of the image; the bottom half is the far side. The two-ring basin Schrödinger (320 km diameter) is at the lower right (about four o'clock). The dark region near the pole indicates an old depression (the pole is just inside the rim of the South Pole–Aitken basin, see Plate 10). This area is probably permanently shadowed, and radar reflectance data from the Clementine mission suggest that deposits of water ice exist here.

and resource potential (ice deposits) ensures its desirability as the site of a future mission. In the longer term, such a locality would be highly desirable as an outpost site for human habitation. With these data from *Clementine,* we may have identified the most valuable piece of real estate in the inner solar system.

Clementine was a small project and thus was able to partly escape the intense scrutiny and sniping to which all space efforts are now subject. It was immensely successful and served as a reminder that interesting, innovative lunar and planetary missions do not have to break the national budget. But what does the Clementine experience tell us about returning to the Moon? Indeed, has it shown us that such a goal is possible? More important, *why* should we want to return to the Moon? And if we do choose to return, *how* should we do it? Much of the first part of this chapter has shown how *not* to return to the Moon. Next let's look at why and how we *should* return.

 Chapter 9

Why?
Four Reasons to Return to the Moon

One of the questions most often heard in regard to a return to the Moon is, Why? Of what use is the Moon? What are the benefits of resuming lunar spaceflight? Why would anyone want to go there anyway? The Moon is a barren, lifeless rock in space. How could we live there?

There are many compelling reasons to return to the Moon. Not the least of these would be to reenergize our space program with a focus, some long-term goal or strategic horizon that has been lacking since the end of the Apollo program. One could argue that another goal, such as a manned Mars mission, would accomplish this praiseworthy task. Instead I will show that there are four principal reasons that make a lunar return unique: the intrinsic value of the Moon as an object of study; our ability to observe the universe around us from the Moon; the chance to use the resources available on the Moon to "bootstrap" a space-faring civilization; and the opportunity to learn about ourselves in the new environment of space. These themes can be addressed *only* on the Moon.

A Natural Laboratory for Planetary Science

From the explorations by Apollo and its robot precursors, we have seen that the geological story of the Moon is much more complicated than was widely thought before the space age. Quite different from the cold, primitive body envisaged by Harold Urey, the Moon is even more complex than imagined by "hot-moon" advocates such as Ralph Baldwin. The Moon underwent near global melting at the time of its creation and experienced over 500 million years of crustal formation, including a variety of melting episodes at different times, to different degrees at different places. On top of this internal activity is a complex history of impact bombardment, a barrage that has broken up, crushed, and mixed the crust. Our understanding of this complexity is the scientific legacy of Apollo, and we would not possess this view without that exploration.

As fascinating as this is, our views on the evolution and history of the Moon have more relevance than just to lunar study. Over the last 20 years, we made our first exploration of the planets. We surveyed and photographed all of the other terrestrial planets—Mercury, Venus, and Mars—and conducted our initial reconnaissance of the rocky and icy satellites of the giant outer planets. We landed robot spacecraft on Mars and analyzed its surface materials. All of the planetary bodies studied to date show, to different degrees, the same kinds of surface and geological processes first recognized and described on the Moon. Much of our understanding of planetary processes and history comes by comparing surface features and environments among the planets. In any such comparison, reference is inevitably made to knowledge we obtained from lunar exploration.

One of the most startling results from Apollo was the concept of the magma ocean (see Chapter 6). The Moon is a relatively small object, transitional in size between the smaller planets and the larger asteroids. In general the amount of heat a planetary object contains is related to its size, with larger planets containing more heat-producing elements. If a body as small as the Moon could undergo global melting, it is a near certainty that the other terrestrial planets melted as well. The idea that the early Earth underwent global melting has been bandied about for many years; the evidence of the magma ocean on the Moon made such speculation respectable. We now think that early planetary melting may have been a widespread phenomenon and could be responsible for the creation of all of the original crusts of the planets.

Knowing that global melting occurred is one thing; understanding, in detail, how it operated is another task altogether. The Moon is a natural laboratory to study this process. One of the most fundamental discoveries of the Clementine mission was that the aluminum-rich, anorthosite crust is indeed global (Plate 9), providing strong support for the magma ocean model. Our next task is to understand the complex processes at work in such an ocean. Did a "chilled" crust form, and if so, are any pieces of it left? Such material would allow us to *directly* determine the bulk composition of the Moon, a parameter that now is estimated indirectly (and very imprecisely). Are there any highland rocks that are highly magnesian and that are related to the magma ocean, not to the younger magnesium-rich suite of rocks? We have searched the Apollo collections for such rocks but have found none. They may exist at unvisited sites on the Moon.

Another process common to all of the planets is volcanism. The Moon is the premier locality to study planetary volcanism. The flood lavas of the maria span more than a billion years of planetary history and probably come from many different depths within the Moon. Thus the lava flows are actually probes of the interior of the Moon, both laterally (across its face) and vertically (through the depths of the mantle). The inventory and study of the mare basalts will allow us to categorize both of these dimensions, with the important additional dimension of *time*. By sampling, chemically analyzing, and dating many different samples of lava that cover the globe, we can piece together the changing conditions of the deep mantle over long times.

The styles of eruption responsible for the maria appear to be typical of those on other planets. Flood volcanism—the very high rates of effusion responsible for the mare lavas—is seen on every terrestrial planet and appears to be especially widespread on Mars and Venus. What is largely unknown is the size and shape of the vents through which these lavas were extruded. On the Moon, eruptive vents might be exposed in several locations, including within the walls of grabens (Fig. 5.10) and irregular source craters (Fig. 3.6). Detailed exploration and study of such features would help us understand a style of volcanism ubiquitous on the planets. Small, dome-like volcanoes, such as found in the Marius Hills (Fig. 5.7), can be explored and examined to understand the styles and rates of lava eruption in the creation of these features. Small shield volcanoes are common on the surfaces of Mars and, particularly, Venus. Relatively exotic processes, such as the erosion of terrain by flowing lava, have been proposed for the sinuous rilles of the Moon. The study of large rilles could help us decide whether this concept is correct.

From the study of the Moon, we know that impact is one of the most fundamental of all geological processes. With its population of craters of all sizes, where better to study and understand this important shaper of surfaces than on the Moon? Our ignorance is particularly vast for craters at the larger end of the size spectrum. Craters such as Copernicus (93 km in diameter) offer a window into the upper crust, through the study of their ejecta; through their central peaks (which have uplifted rocks from 15 to 20 km deep, Fig. 9.1), they provide a view of the middle level

Figure 9.1. Close-up view of the central peak of the crater Copernicus (see Fig. 2.19). From study of impact craters on Earth, we know that central peaks are uplifted from very deep levels of the crust. Thus the study of central peaks on the Moon will allow us to look at and analyze the deep crust.

of the crust. The ubiquity of craters with such central peaks allows us to reconstruct the nature of the crust in detail. The study of large craters will also clarify the nature of the process of impact. We suspect that large craters grow proportionally, this is, they excavate amounts of material to depths that can be predicted from the study of smaller craters. But we are not certain this is true. By studying craters on the Moon, we can determine whether this pattern of growth behaves as predicted.

The giant basins of the Moon pose many mysteries. Understanding such craters is important because we have found basins on the other planets, particularly on Mars and Mercury. Basins form in the earliest stages of planetary history. They excavate and redistribute the crust, serve as depressions where other geological units may be deposited (such as stacks of thick lava), and may trigger the eruption of massive floods of lava. Yet for all of their importance, we still do not fully understand how far or how deep excavation extends, how the multiple, concentric rings are formed, how the ejected material behaves, and how far ejecta gets thrown as a function of mass. The Moon has preserved over 40 of these important features for our study, all in various states of preservation and all dating from the very earliest phase of planetary history. Here we can study the process of large-body impact better than anywhere else in the solar system.

The temporal record of impacts in the Earth-Moon system can also be read on the lunar surface. On the basis of evidence for mass extinctions on Earth and from the ages of impact melts, cycles of bombardment (see Chapter 4) and an early impact "cataclysm" (see Chapter 6) have been proposed. Neither of these ideas has been proven, but both are potentially revolutionary in nature, causing us to look at the history of the planets in a new way. The evidence to test these two ideas lies on the Moon. Episodic bombardment can be tested by sampling the melt sheets of many different craters (Fig. 9.2) and dating the samples. Episodes of intense cratering will be evident if groups of melt rocks have the same ages, spaced at constant intervals. On the other hand, a continuous distribution of ages would argue that such episodes of intense cratering do not occur. We cannot conduct this experiment on Earth or on any other planet, a fact that

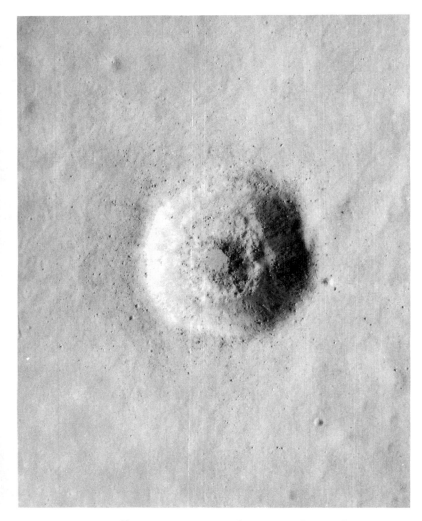

Figure 9.2. A small impact crater on the Moon, about 1 km in diameter. The smooth pool of material in the center is impact melt produced during crater formation. Sampling and dating rocks from a number of such melt sheets will allow us to determine precisely the cratering history of the Earth-Moon system.

highlights the uniqueness of the Moon for answering many questions in planetary science, questions that pertain to a host of other scientific fields. The cataclysm is important because if the Moon underwent such an unusual bombardment history, Earth may also have experienced it; thus the process of applying the inferences made from lunar cratering to the other terrestrial planets would have to be reevaluated.

The study of the regolith will be one of the most important tasks during a lunar return. The regolith contains exotic samples flung from rock units hundreds of kilometers away. Using the regolith as a sampling tool, we can conduct a comprehensive inventory of the regional rock units by collecting many samples from the regolith at a single site. The regolith also contains a record of the output of the Sun over the last 3 billion years. To read this record, we must understand how the regolith grows and evolves. This knowledge will come only when we are able to study the regolith and its underlying bedrock in detail to learn how its layers are formed, how the soil is exposed, buried, and reexposed, and how volatile components might be mobilized and migrate through the soil. This knowledge is essential if we are to realize the goal of using the regolith as a recorder of the solar and galactic particles that have struck the Moon over its history.

We study the geological processes and history of the Moon because they are well exposed here and because they can be easily reached. The Moon is a natural laboratory for planetary science. What's more, because it is only three days away by spacecraft, this lab is right next door!

A Platform to Observe the Universe

The Moon will one day become humanity's premier astronomical observing facility. Consider its advantages. The Moon rotates very slowly (once every 29.5 days), so its "nighttime" is two weeks long. Moreover, because the Moon has no atmosphere, we can observe stars constantly, even in the daytime! The lack of an atmosphere also means that telescopes on the Moon will not be plagued by the blurring caused by a turbulent, thermally unstable air layer and that observations will not be degraded by the light pollution and the airglow that interfere with astronomy on Earth. The vacuum of the Moon also means that there are no absorptions to prevent certain wavelengths of radiation from being observed, such as the infamous "water absorption" in the atmosphere of Earth. The surface materials could be used as construction material for observatories.

Telescopes in Earth orbit or elsewhere in deep space also have many of these advantages. Why is the Moon better than these localities? The principal reason is that the Moon provides a quiet, stable platform. Seismic activity on the Moon is roughly one million times less than on Earth. Because the Moon is a primitive, geologically dead world, it does not have the shifting, massive plates of our own dynamic Earth, along with its associated seismic trembling. Such stability of the surface would allow us to construct extremely sensitive instruments for observing, ones that could not be constructed on Earth.

Likewise, a space-based telescope, such as the Hubble Space Telescope, must have its attitude carefully stabilized in order to achieve high-resolution observations. In addition, space telescopes have stringent pointing requirements and must not be pointed anywhere near the Sun. Both of these drawbacks mean that a telescope free-flying in space must carry attitude-control fuel, precision gyroscopes, and equipment to protect the telescope optics from solar burn damage. On the Moon the quiet, stable base of the surface would alleviate such problems, allowing sensitive instruments to be erected and operated easily.

One such instrument, an interferometer, consists of an array of smaller telescopes (Fig. 9.3). The smallest object that a telescope can see clearly is directly related to the size of its *aperture*, or the diameter of its mirror or lens. A telescope with a larger aperture can resolve smaller or more distant features than can a telescope with a smaller aperture. However, there is a practical limit to the size of telescopes; after a certain size is attained, such instruments become unwieldy and unstable. Interferometry is a technique whereby a series of small telescopes are operated as a larger-aperture instrument. Each element of the array images some distant object. The light waves from this image are "added" in perfect phase and frequency to identical images obtained from other telescopes in the array, each separated by as much as sev-

Figure 9.3. An interferometer, an array of small telescopes that can be operated on the Moon as a single instrument. In such an arrangement, the interferometer has an effective aperture equal to the separation of the individual telescopes, which can be kilometers. Such arrays will allow us to see the universe in unprecedented detail. Illustration courtesy of EagleAerospace, Inc.

eral kilometers. The effect of this addition is to create an image of the same quality that would be produced if a telescope with an aperture size equal to the separation distance had been used to image the star. This means that we can construct "telescopes" whose effective aperture sizes are *kilometers* across!

Even a small interferometer on the Moon would exceed the resolving capabilities of the very best existing telescopes on Earth and would even surpass the capabilities of the Hubble Space Telescope. With such an instrument, we could resolve the disks of distant stars, to observe and catalog "star spots," which are clues to the internal workings of stars. We would be able to see individual stars in distant galaxies and catalog the stellar makeup of a variety of galaxy types. Optical interferometers would be able to look into a variety of nebulae and observe the details of new stars and stellar systems in the very act of formation. Such a window onto the universe would very likely revolutionize astronomy in the same way it was changed when Galileo turned his crude "spyglass" toward the heavens back in 1610.

The field of planetary astronomy would be completely changed by lunar observatories. The incredible resolving power of these instruments would allow us to examine deep sky systems, resolve the disks of extrasolar planets, and catalog the variety possible in other solar systems of our galaxy. All of our concepts of how planets are created and evolve are derived from a single example, our own solar system. By observing the vast array of planetary systems circling nearby stars, we would be able to see how they differ in such aspects as the number and spacing of planets, the ratio of giant gas planets to rocky "terrestrial" objects, and the evolution of those individual planets. Spectroscopic observations of these planets would allow us to determine the composition of their atmospheres, if any, and the surface composition, if visible. The composition of planetary atmospheres could indicate the presence of life on these bodies. Carbon dioxide mixed with free oxygen in a planetary atmosphere would be a telltale indication of plant life, which "breathes" the former and manufactures the latter.

Astronomers look at the sky in many wavelengths other than the optical band. High-energy regions of the spectrum, such as X-ray and gamma-ray radiation (Fig. 8.2), also contain important information about processes that occur in stars and galaxies. Supernovae (the sudden explosion of certain stars) produce copious amounts of high-energy radiation and energetic particles, such as cosmic rays. We think that certain particles derived from supernova explosions predate our solar system and that these star eruptions can induce planetary formation. We have already mentioned the use of the regolith as a recorder of energetic particle events. With a high-energy lunar observatory, we can watch the effusions of these particles and radiation as they happen (supernovae are common and typically are occurring in some part of the sky at any given time). We have only begun to observe such stellar explosions from space and no doubt have much to learn.

At the other end of the spectrum, the Moon is an ideal place to observe in the thermal infrared and radio bands. Observing the sky in the long wavelengths of the thermal infrared (10 microns and longer wavelengths, Fig. 8.2) is difficult because such detectors measure heat and they must be cooled to very low temperatures for use. Usually, this is done at great cost and difficulty with cryogenic gases, such as very cold liquid helium (−269°C).

Preserving such cold temperatures requires a lot of electrical power. The Moon is naturally cold. Surface temperatures during the lunar night may reach as low as −160°C. Shadowed areas near the south pole (Fig. 8.6) may be as cold as −230°C, only 40° above absolute zero. These temperatures would permit passive cooling of infrared detectors, allowing telescopes to be operated without costly and difficult-to-use cryogenic cooling gas. Observing the thermal infrared sky would tell us much about dust clouds and nebulae in which new stars and planets are being formed.

The sky at certain radio wavelengths is almost completely unknown. Earth has an ionosphere, a layer of electrically charged atoms that causes certain radio waves to bounce off it, and we cannot see the radio sky at certain frequencies. Moreover, the electrical din of Earth caused by radio stations, microwave cookers, automotive ignition systems, and the thousands of other static generators of modern civilization vexes radio astronomers in their attempts to map the sky. The far side of the Moon is the *only* known place in the universe that is permanently shielded from the radio noise of Earth. Locating a radio telescope on the far side would permanently place 3,600 km of solid rock between the observatory and the radio din of Earth. We will see sky for the first time at some radio wavelengths. History has shown us that whenever we look at the universe with a new tool or through a new window of frequencies, we learn new things and reexamine present knowledge with new and sometimes startlingly different appreciation.

We can use the distinctive lunar terrain to our advantage. The small, bowl-shaped craters of the Moon are natural features that could be turned into gigantic "dish" radio antennas (Fig. 9.4) by laying conductive material (e.g., chicken wire) on their floors and hanging a receiver over and above the center of the crater at the "focus" of the dish. This technique has already been done on Earth at the famous Arecibo Radio Observatory in Puerto Rico, using a natural depression in the limestone bedrock to create a giant dish antenna. Interferometers could also be built at radio wavelengths, creating radio telescopes with huge apertures. The large, flat mare plains would make an ideal site to lay out arrays of smaller telescopes. The manufacture of antenna elements

Figure 9.4. A lunar crater being used as a radio telescope. We can use the natural topography of the lunar surface to construct astronomical instruments. In this case a small, bowl-shaped crater is lined with mesh to form a large, fixed-dish antenna for a radiotelescope.

from the local resources could make the construction of extremely large instruments feasible.

Using the Moon as an astronomical observatory has great advantages, and many astronomers have taken up the banner for a return to the Moon. In the minds of some, astronomy is the principal reason for a lunar return. However, an observatory on the Moon also has its problems. The ubiquitous and highly abrasive dust must be very carefully controlled. Movements of people and machines will have to be minimized around telescope facilities because the slightest stirring up of dust could coat delicate optical surfaces. We would have to carefully shield energetic detectors from solar flares (this could be done by using the local regolith material). We must guard against the radio contamination of the far side because extensive operations associated with a base could ruin certain radio astronomical observations. Our task before a lunar return is to fully understand the effects of each problem and to devise methods of working around them.

Despite these problems, the Moon offers unique opportunities for astronomy. Each time we see the sky more clearly or more completely, we obtain new insights into the way the universe works. A lunar window on the universe around us will give us a new appreciation and understanding both of the universe and of our place in it. Because such a new view is likely to be completely revolutionary, no one can truly say what we will discover. If history teaches anything, we can say one thing about the new knowledge obtained with the unobstructed view from the Moon's surface: It will be astonishing!

A Source of Materials and Energy for Use in Space and on Earth

Let's dispose of one concept right away: There are no massive gold deposits, no huge diamond mines, no elephant burial grounds of ivory waiting to be picked up, no El Dorado on the Moon. Some of the wilder pre-Apollo speculations about the Moon, suggesting incredible mineral deposits, have been refuted. The Moon is rather ordinary in composition. The aluminum-rich crustal rocks are made up of plagioclase feldspar, a very common mineral on Earth—so common in fact that the principal economic use of feldspar in Earth products is as an abrasive in kitchen scouring powder (Ajax—not the Hope Diamond!). The basalts of the maria are similar to very common rocks on Earth, where they are used primarily as building stone. The total absence of water in the melting episodes that occurred early in lunar history ensured that massive hydrothermal ore deposits. the type that produce much of Earth's mineral wealth, are completely absent from the Moon.

Nevertheless, the Moon is an object of incredible potential wealth. What exactly do we mean by the term "ore deposit" anyway? Simply put, an ore is a deposit that can be mined at a profit. Thus ore is an economic term, not a geological one. The word *profit* perhaps requires some rethinking. It does not necessarily mean money. In the field of space resources, profit could mean the ability to use something found on another planetary surface instead of carrying the resource from Earth. It costs many thousands of dollars to lift water, oxygen, and rocket fuel from Earth's surface to orbit. If these materials are already in orbit, we can save a *lot* of money. The Moon is already in Earth orbit.

Resources have two aspects: materials and energy. Materials such as water, oxygen, and other elements can be used for supporting human life, for building structures in space, and for supplying rocket fuel. Fuel is also an energy source. For example, if oxygen and hydrogen are produced on the lunar surface, they can be used to generate electricity in machines called *fuel cells*. Solar energy has been used for electrical and thermal power since we have flown in space. The two-week daytime of the Moon permits us to use solar energy continuously for periods as long as 350 hours. The Moon offers abundant resources for our use, if we are clever enough to learn how to use them.

The bulk soil offers a ready resource, and it is likely to be one of the first materials mined and utilized on the Moon. Because there is no global magnetic field, cosmic rays and solar flares continually hit the surface, making it a very harsh radiation environment. If people are to live on the Moon for any extended period, they must have protection from this radiation. One of the simplest expedients is to cover surface habitation modules

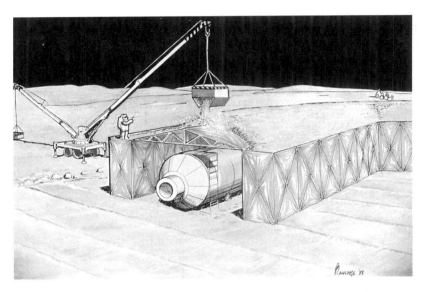

Figure 9.5. Soil being shoveled onto prefabricated living modules carried from Earth to create the initial outpost. One of the first uses of regolith will be to provide shielding from radiation and micrometeorites for lunar inhabitants.

with loose soil (Fig. 9.5). A covering of about 3 m of regolith on the habitat would protect the inhabitants from most of the harmful radiation. In practice, habitats would be placed on the surface, and then a loader or shovel would cover the modules with soil. More sophisticated shielding can be made by melting the loose dirt into blocks that could then be used as building bricks for more advanced base elements. Such soil sintering could be done easily in solar furnaces, made by using a simple, concentrating mirror.

The soil of the Moon can also be used to make glasses and ceramics. Microwave ovens have been used on Earth to melt simulated lunar soil into glass. Because real lunar soil contains no water, glass made from such material is extremely strong and can be used for many building tasks, including support beams for buildings. Thus glass made from processed soil may be the steel and concrete used to build future lunar cities. By melting the regolith in place, we can "pave" the routes used to service the outpost and distant facilities, allowing us to make roads for vehicles and keeping sprayed dust from covering everything. Likewise, ceramic materials made from soil have many industrial uses, including electrical insulators, thermal shields, and aerobrakes, large, flat ceramic disks that can be used as heat shields to permit spacecraft returning from the Moon to slow down into Earth orbit without burning rocket propellant.

The Moon has no air, but surface materials consist of about 45 percent oxygen by weight. The problem is that these oxygen atoms are held in very tight chemical bonds with silicon, aluminum, iron, and other elements. These bonds can be broken but only at great cost in terms of energy. The trick to the production of lunar oxygen is to find the easiest way to extract the gas from its host material. Most of the processes investigated to date rely on chemical reactions that liberate oxygen from these compounds. In one process, hydrogen (carried to the Moon from Earth) is blown across a heated bed of titanium-rich mare regolith. The hydrogen reduces the iron-titanium mineral ilmenite, and the chemically released oxygen combines with some of the hydrogen to make water vapor. This vapor is then captured, cooled, and stored as liquid water. The ilmenite reduction process has an efficiency of only a few percent but is relatively easy to undertake and is fairly well understood. In another process, regolith is completely melted, and an electric current is passed through the melt. This technique, called *magma electrolysis,* has relatively high product yields (up to the full amount of bound oxygen), but a great deal of energy is required to melt the soil completely.

A variety of techniques can be used to extract oxygen from the Moon. For rocket fuel this is good news because 80 percent of the mass of the fuel carried by an oxygen-hydrogen rocket is liquid oxygen. However, it is highly desirable to be able to extract the hydrogen as well, not only because it is needed for fuel but because water will be a prime requirement for living on the Moon. We have seen that water is completely absent from the lunar rocks, but solar wind hydrogen *is* implanted on the grains of dust. Can we "mine" the lunar soil for hydrogen? Hydrogen is relatively rare in the regolith but occurs at concentrations between 25 and 100 ppm in mature, titanium-rich soils from the maria, particularly in the finest grain sizes (less than 50 microns). What's more, this hydrogen is easily extracted; heating the soil to about 700°C drives off most of the solar wind gases. These gases include hydrogen, helium, nitrogen, and sulfur. Such temperatures are easily achieved by solar thermal heating simply by using a focusing mirror. The gas released from the soil can be cooled cryogenically and stored as a liquid.

Volatile elements are important materials for use on the Moon. Solar wind gases are found everywhere but vary in concentration. Hydrogen seems to be particularly abundant in soils developed on high-titanium mare basalt flows. We think that this is because the mineral ilmenite has a crystal structure that acts like a sponge to soak up and retain excess hydrogen. Some deposits also have relatively large amounts of indigenous volatile elements. Dark, pyroclastic glasses (Fig. 5.3) are very fine grained, have high titanium, and are regional in extent. In addition to the high abundance of solar wind hydrogen contained in these deposits, volatile elements such as zinc, lead, and fluorine coat the surfaces of the small glass spheres. All of these elements are industrially useful, and the fine grain size of the regional dark mantle deposits would make them easy to mine.

An operation to produce both hydrogen and oxygen on the lunar surface would be the first real milestone in the use of extraterrestrial resources (Fig. 9.6). The products of this operation will allow us to easily travel between Earth and the Moon. Because landers would not have to carry their own return fuel, they could be made smaller and lighter, significantly reducing transport costs. Any operation producing enough oxygen and hydrogen for rocket propellant would make enough water to support a rather large human outpost. Moreover, the by-products from hydrogen extraction, such as nitrogen and helium, would be useful to a lunar facility. For example, solar wind nitrogen and indigenous phosphorus (the "P" in KREEP) would be essential for lunar agriculture. Understandably, a lot of work has been expended on schemes to "mine" the Moon for its hydrogen and oxygen.

Energy is the other side of the resource coin. Any significant installation on the Moon will require a great deal of energy. In the early stages of lunar growth, we will utilize the abundant solar energy available on the Moon; after all, the Sun shines 14 days at a time—and there are no clouds! The first outposts will

Figure 9.6. A turnkey oxygen-production station. The production of oxygen and hydrogen from the lunar soil would greatly reduce the costs of the Earth-Moon transportation system. Such an extraction system could be started at a very modest level and built up gradually as it pays for itself.

bring up solar-cell panels from Earth, but almost all of the material that we need to make solar cells already exists on the Moon. Amorphous (i.e., glass) silicon cells can be manufactured on the Moon; the minute amounts of trace elements needed to turn them into electric cells can be brought from Earth. Parts of the smooth, flat mare plains could be literally turned into solar panels by melting the soil and making the cells in place, manufacturing the arrays on the surface of the Moon. Very large solar collectors can be built in this manner, each one many kilometers across, allowing us to create huge, multimegawatt power stations without the high costs of transporting the solar arrays from Earth.

Such huge power stations (Fig. 9.7) could supply the needs of a significant colony, at least in the lunar daytime (14 Earth days). Operating the outpost at night (also 14 Earth days) would

Figure 9.7. A solar power station on the Moon. By melting and processing lunar soil in place, we can manufacture solar cells and arrays in very large sizes. In effect, we can turn large parts of the Moon into giant solar power stations. The electricity produced can be beamed by microwave or laser to other places on the Moon or to Earth. Courtesy of David Criswell.

require either fuel cells (which would need hydrogen and oxygen for fuel) or energy carried in from elsewhere. One possibility is to build another power station antipodal (on the opposite side of the Moon) from the main outpost. It would be daytime at the second power station while it is night at the outpost. The energy produced by this installation could be beamed by laser or microwave to orbiting relay stations and, from there, to the other side of the Moon. In this manner power stations could produce electrical energy for the entire month-long lunar day-night cycle.

With a productive global power system, the energy produced would not have to be used solely on the Moon. Clearly, if we can beam power through space with microwaves or lasers, we can send such energy back to Earth. The concept of large, solarpower satellites providing energy to remote localities on Earth is not new. These satellites could orbit Earth in perpetual sunlight, use huge solar panels to turn the light into electrical energy, and beam the power to Earth. The problem with this idea is the huge cost of launching the immense panels into orbit from Earth's surface. But because the Moon is already in Earth orbit, making power stations on the Moon out of lunar materials could solve this problem. It would also have the by-product of making the habitation of the Moon much easier. Some scientists have advocated energy production as the main rationale for a lunar return.

Another scheme to bring energy home to Earth from the Moon relies on the rare gas ^{3}He (helium-3). As discussed in Chapter 4, ^{3}He is rare on the Moon, the highest concentrations being about five parts per billion. However, this concentration is orders of magnitude greater than any source of ^{3}He on Earth. Although fusion reactors that can "burn" ^{3}He do not exist today, they are foreseeable developments in the middle of the next century. Mining the Moon for ^{3}He would be a massive undertaking, requiring a high level of industrial development, but no greater than would be needed for the harvesting of solar energy as described above. In short, the mining of the Moon for ^{3}He is an intriguing idea but one to consider only after we already have our "foothold" on the Moon.

So the Moon is not a jewel box in space or an astronautical El Dorado, as perhaps envisioned by the early lunar "conquista dors." But lunar resources have great potential value. By not only yielding provisions for our journey into the solar system but also teaching us how to obtain what we need and want from low-grade ores, the resources of the Moon can allow us to bootstrap a space-faring civilization. The energy that we produce on the Moon could be used on Earth to lessen our dependence on dwindling fossil fuels and to help industrialize the Third World. Just as important, those materials and that energy will allow us to live and work on the Moon and give us our first sustained step into the solar system. If we can live off the land on the Moon, we can live anywhere in the solar system.

A Place to Learn to Live and Work in Space

The Moon is a natural space station, orbiting Earth every 27 days. In spaceflight terms it is very close, typically a three-day journey. Although not as easy to reach as low Earth orbit, it is far easier to get to than any of the other planets and most of the asteroids. Earth remains reassuringly overhead, only 1.5 light-seconds away by radio. Yet at the same time, the Moon is a world unto itself, offering mysteries worthy of a great exploratory impulse. If people are to have a future in space, we first must learn how to operate in this new and unfamiliar environment. The Moon holds many opportunities for learning and is near enough to make this first great step a relatively painless one.

The most vexing questions about humanity's future in space are those about people themselves and their adaptation and reaction to space. Much is made about the possible ill effects of long-term weightlessness on the human body. After long exposure to *zero-g* (*microgravity* is the new, approved buzzword), there is permanent calcium loss in bones and atrophy of the muscles. So much concern has been raised over this issue that some scientists have questioned whether long space travel, such as would be required for a Mars mission, is even possible. I do not want to minimize the physiological problems of zero-g, but we already know how to mitigate this problem. We can either shorten the trip time (through the use of nuclear rocket engines, which can cut travel times in half) or spin the spacecraft to create artificial gravity for long periods (as shown in the movie *2001: A Space Odyssey* with the spinning crew quarters). In other words, the *physiological* problems of long-duration spaceflight can be addressed and solved.

There is a much more fundamental and important issue at the core of our assumptions about human spaceflight. Even after purely physical problems, such as adaptation to low gravity, are overcome or alleviated, can humans effectively live and productively work in space for very long periods? This remains to be determined. Our limited experience with long-duration flight (with the Skylab and Mir missions) and other high-isolation environments (such as polar bases) shows that productivity, alertness, and capabilities suffer over very long periods. People get complacent and sloppy. The lack of privacy and the constant direction from ground control begin to irritate, and depression sets in. It is even possible for sociopathic behavior to develop, a disaster in a situation in which close cooperation is essential. Clearly, such problems can be addressed and controlled for limited periods of time, when the mission is defined and the end is well in sight, such as in the Apollo missions or possibly even in a future mission to Mars. But will the inhabitants of a lunar base also be as motivated and productive?

I suggest that these problems are vitally important to the future of humanity in space and are addressable by living and working on the Moon. In this sense, the Moon can be looked at as a psychological and sociological test laboratory as well as a physiological one. If people are to be useful and productive in space, we must be able to successfully settle and work on the Moon. Because the Moon is close, potential problems can be recognized early and experiments terminated easily, if need be. The physical nearness of Earth, always overhead in the same place on the near side, would be immensely reassuring psychologically. However, when privacy is desired, the far side is as physically isolated as one can get and still remain within the solar system. Even direct radio contact is not possible.

How will we live off-planet? Early habitats are likely to be fabricated on Earth and brought up, ready to use (a turnkey operation). As the outpost grows first into a base, then into lunar towns and cities (Fig. 9.8), construction from local materials will

Figure 9.8. A city on the Moon. Ultimately, through the use of indigenous resources, the lunar colony will grow into large towns and cities. What kind of social organization will the new-worlders require?

be necessary. How will this be done? What materials are easiest or best to use? (The two may not be the same.) Some sinuous rilles in the maria may once have been roofed-over lava tubes (Fig. 5.8), and it is not impossible that caves exist on the Moon today. If found, such lava tubes would be very attractive habitats, offering a safe and secure shelter from the harsh radiation at the surface and possessing a constant, benign thermal environment. All we would need to do is seal the cave and pressurize it with an atmosphere. Can large numbers of people live underground for long periods? We can answer these questions by living on the Moon.

A related but distinct set of questions involves strategies of space operations. There has always been a debate about the relative merits of people and robots as space explorers and workers. Some scientists see little value in having people in space, whereas others advocate a balanced program with both human and robotic explorers. The Moon once again is the ideal test laboratory in which to understand the trade-offs and merits of each approach and, more interesting, to see how people and machines can work together in a symbiotic, beneficial arrangement.

One aspect of such a cooperative mode might be the use of teleoperations. In this technique, a robot is remotely located in an area hazardous for human occupation, for example, the lunar surface. This machine has two robotic eyes, two arms, hand manipulators and fingers, and some means of locomotion (Fig. 9.9). The robot is *not* autonomous but is under the control of a human operator, who is safe in the habitat on the Moon or in orbit around the Moon. Because the robot "sees" in stereo and because a sense of touch is transmitted by tactile feedback sensors, the human operator has the sensation of being "present" at the remote location, of being inside the robot. In effect, this technique allows people to roam over the lunar surface, exploring and discovering at will, without the need for life support and without the thermal and radiation hazards of extended exposure on the surface.

There is no doubt that technically such a machine can be built. But will this experience be the "same" as actually walking on the Moon? Can the explorations and scientific investigations done by the telerobot be as efficient and as thorough as those done by a

Figure 9.9. A teleoperated robot conducting geological exploration on the Moon. In this technique, sensory data from the robot are sent to an operator at a remote location, providing the illusion that the operator is present at the remote site. We need to conduct experiments to determine whether this approach can be as productive as having people do the actual surface exploration.

human explorer? If the operator is separated from the robot by great distances (for example, if the operator is on Earth), a significant time delay may exist while radio makes its round-trip (from Earth to the Moon, 2.8 seconds). Will such a delay degrade the telepresence effect? Can work be done efficiently under such a constraint? In some cases we can use robotic sensors to *increase* the sensory capacity of the human operator by giving the robot multispectral vision, making it sensitive in regions of the spectrum where humans cannot see. Does this compensate for the sensory degradation of not being "there"? Or are we really in fact "there," remotely present in all respects?

Such a potentially useful tool for exploration could be carefully evaluated in the course of a return to the Moon. If successful, this technique could be applied to other planetary missions, including trips to localities that humans cannot travel to, such as the surface of Venus or the atmosphere of Jupiter. A related issue is the desirable interplay of robotic and human missions. In the past, people have followed where robots blazed a trail. Is this the only possibility? Perhaps having follow-up visits by robots to sites already explored by humans could materially advance our understanding of the processes and history of the Moon. In such a strategy, what is the optimum mix of human and robotic missions? Using the Moon as our laboratory, we can learn the answers to these questions.

On a broader level is the issue of society in space. If we can settle the Moon, extract what we need from the local resources, and create products for markets in space and back on Earth, what kind of social organization does that imply? Will self-sufficient lunar inhabitants demand independence? If so, what is their obligation to Earth? What are our obligations to them? Such questions have littered history with more than a few wars (and at least some periods of extreme nastiness!). Will an insular, self-sufficient colony on the Moon thrive? Will it self-destruct? Or will humanity's outpost on the Moon remain an isolated duty station, a research base, similar to those of Antarctica?

Each question prompts many more questions. Yet the questions raised are of fundamental significance to our future life off-planet, beyond Earth. We are truly fortunate to be endowed with such a conveniently located, fascinating world—a place to explore, to understand, to use, and to inhabit. On the Moon, we can learn if mankind has what it takes to settle the solar system and to move out beyond it, into the stars.

 Chapter 10

How?
Steps in the Exploration and Use of the Moon

Once we decide that one or more of the reasons for a return to the Moon are compelling, we will face a problem similar to that faced by the United States over 30 years ago: How do we go to the Moon? I do not mean by this that we need to determine the type of rockets, the masses, or the escape velocities required for space travel. All of these factors are well understood and are dictated primarily by physical laws and the engineering state of the art. I mean that we should discuss what missions to undertake, which techniques to use, and in what order to conduct the missions in a return to the Moon.

After a decade of study, we understand the options involved in constructing a return to the Moon. These options concern more than the traditional (and pointless) debate about the relative values of "manned" and "unmanned" spaceflight. We must think anew about how robots and people can work together to accomplish much more than operating either separately or alone. We must distinguish between the *required* steps and the merely desirable ones. Our strategy for return determines the scientific and exploratory direction of the lunar establishment. An architecture may be tailored to emphasize specific research priorities as needed.

A return to the Moon is scalable to the resources available. One of the arguments used to torpedo the Space Exploration Initiative was that it was a "$500 billion, 30-year program" (see Chapter 8). This was and is a foolish argument against a return to the Moon. Apart from an appeal to religion or national defense, there has never been another program conceived and sustained on such a scale for so long a time. A lunar return can and should consist of a series of steps that build upon each other. Each step is relatively small and stands on its merits. Moreover, each step or "phase" should have a time horizon no greater than about five years; a program much longer is difficult to sustain politically. A lunar program is like a bolt of cloth; one can "buy it by the yard." With this approach, we can even *make* a little money on the deal, if we are clever enough.

Orbital Surveys: Clementine and Future Missions

A global survey of the Moon from orbit is an important step toward a lunar return. To assure ourselves that we are going to the most operationally and scientifically desirable location, we should map the Moon globally in a variety of wavelengths and at high resolutions for certain data. This relatively inexpensive step would alleviate concerns for safety or mission integrity to certify (examine in detail) the base site before we send people there.

The data provided by the Clementine mission go a long way toward satisfying these needs. From Clementine, we have complete coverage of the Moon in 11 key wavelengths in the visible and near-infrared parts of the spectrum. From these data, we will be able to map minerals and rock types and their distribution across the surface and at depth. For the first time, we have a map of the shape (or topography) of the Moon, allowing us to address its thermal history and the processes involved in the formation of large impact basins. Clementine also provided high-resolution color images and thermal infrared data for large areas of the surface.

There are still many gaps in our knowledge that could be filled by additional orbital missions. The most pressing need is for global maps of the chemical composition of the surface. From Clementine data, we can map the distribution of rock types on the Moon, but some very different rock types can have identical mineral contents. For example, a common rock from the highlands is norite, which contains more or less equal amounts of plagioclase and the magnesium-rich silicate pyroxene. Norites are common in impact-melt rocks and in the ejecta of some basins, such as the Serenitatis basin sampled by the *Apollo 17* mission (see Chapter

6). However, the component KREEP is also typically found in norites. Some norites have very large amounts of KREEP, whereas others (the Mg-suite norites) have much less, and each type of rock has very different implications for the composition and history of the crust in the regions it is found.

Although the multispectral data from Clementine can map the distribution of norite, it cannot distinguish between KREEP-norite and Mg-norite. We can discriminate between the two with a map of the *chemical* composition of the surface because KREEP-norite has very high amounts of the radioactive elements thorium and uranium. We need to know the chemical composition of the entire surface to fully assess the distribution of other rock types as well. Moreover, spectral data can detect glass but can yield only limited information about its chemical composition. Because glass makes up much of the regolith, maps showing the chemistry of the Moon will allow us to fully assess the composition of large regions.

Many other wavelengths have not been examined from orbit. Clementine gave us our first close-up look at the Moon in the thermal infrared. This part of the spectrum contains important information on the chemical and mineral makeup of surface materials. Mapping the surface at many wavelengths in the thermal infrared would provide new information on rock types and act as a cross-checking technique for the maps derived from other techniques. Additional sensing both at longer wavelengths (thermal microwaves) and at shorter ones (ultraviolet and X-ray) can also provide additional information on the distribution of materials on the surface (Fig. 8.2).

Compositional mapping is not the only task for orbital surveys. We still need detailed maps of the gravity and magnetic anomalies of the Moon. Even after Apollo and Clementine, we lack a complete set of surface images taken at low to moderate sun angle illumination to make geological maps (as described in Chapter 2). From such data, we can also map surface topography with stereo coverage at finer scales than the Clementine mission could provide from laser altimetry. Such detailed mapping is needed at least for sites that we may want to inhabit (the base site), build upon (sites for an astronomical observatory), or explore (sites of high geological interest).

The technology of remote sensing constantly advances, and there will always be a need to fly orbital missions with new, sensitive detectors. In addition to passively measuring the reflected radiation of the Moon, there are also "active" sensing techniques. Mapping from orbit by radar can "see" into the permanently shadowed regions of the south pole (compare Figs. 8.6 and 10.1). The Soviet Phobos mission in 1989 to one of the small moons of Mars planned to fire a laser beam at the moon, vaporizing the soil in a brief flash. This glow could then have been analyzed to get chemical information about the surface of Phobos (unfortunately, this mission failed after communication with the spacecraft was lost). Such "active" sensing techniques could also be applied to the Moon from orbit. One of the most promising ideas is to use a stream of high-energy neutrons, produced by a nuclear reactor in lunar orbit, and beam these particles to small, focused spots (1–2 km across) on the Moon. A gamma-ray detector could then measure the complete major- and trace-element composition of these spots. Through this technique, we can remotely map the chemical composition of sites at much greater resolution (1 km) than is obtainable by passive gamma-ray mapping, which can resolve features only hundreds of kilometers across.

Orbital missions are needed both now and later, after the lunar base is established. Such missions can conduct surveys at many scales, from local to regional to global. They can carry new and more advanced sensors and techniques to provide different information. Such missions provide data for strategic planning, such as searching for important scientific sites, and for tactical operations, such as planning base or facility installations. Orbital missions are an important tool in the exploration and use of the Moon.

Fixed Stations: Robot Telescopes and Observatories

The installation of fixed stations on the lunar surface is another important class of mission. Some exploratory and scientific tasks require instruments or facilities on the Moon, but mobility is either unnecessary or undesirable; observatories have such characteristics. On the Moon we can look out, at the uni-

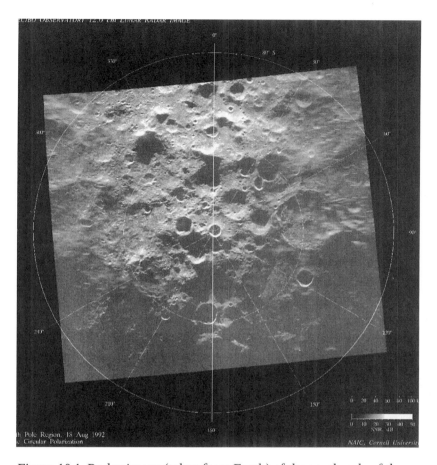

Figure 10.1. Radar image (taken from Earth) of the south pole of the Moon. Compare this view with the visible-light images of Fig. 8.6; with radar imaging, we can "see" into permanently dark areas. On future orbital missions, various advanced remote-sensing techniques can be used to learn new things about the Moon. Courtesy of Don Campbell.

verse around us, or look down, into the very heart of the Moon itself.

One of the astronomical instruments likely to be placed on the Moon is a fixed, large-mirror (about 1 m across) telescope. This instrument, called a *transit telescope,* could be placed at a site near the equator and would use the motion of the Moon itself to survey the sky. As the Moon slowly rotates on its axis and revolves around the Sun, the entire equatorial region of the sky would come within the field of view of the telescope. In such a manner a nearly full sky map could be made at high resolution and complete spectral coverage, without the complexity of steering motors and heavy, precision telescope mounts. Ultimately, steerable telescopes will also be built. These facilities will require major efforts to locate suitable sites and to build a supporting infrastructure around the observatory, including a road network for vehicles to service the telescopes (Fig. 10.2).

Lunar observatories can also be used to observe Earth and the Sun and to monitor the near-Earth *magnetosphere*, the large regions of space containing electrically charged particles (Fig. 10.3). Fixed instruments can provide constant "snapshots" of current conditions, much like the network of weather satellites in geostationary orbit. High-energy particles from the Sun and galaxy can be monitored and observed from fixed stations, which will allow us to better understand and recover the record

Figure 10.2. Astronauts servicing an astronomical observatory on the Moon. Unlike robots flown in space to date, humans have a unique ability to service and maintain complex equipment. Routine and emergency maintenance will likely require constant human intervention. Illustration courtesy of Lockheed Missiles & Space Co.

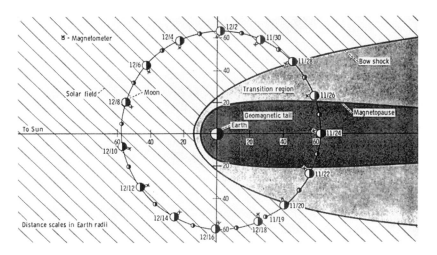

Figure 10.3. The magnetosphere of the Earth-Moon system. As the Moon orbits Earth, it moves in and out of Earth's magnetosphere, the zone of magnetic field lines and solar charged particles that is partly responsible for weather, radio propagation, and aurora (the "northern lights"). We can use the Moon as a platform to monitor the magnetosphere.

of these radiation events preserved in the regolith (see Chapter 4). We hope to observe the elusive "gravity waves" postulated by theory, which we first attempted to observe in 1972 from the Moon on the *Apollo 17* mission (the experiment would not "unlock" itself and was thus a failure). Gravity waves could offer clues to the origin and makeup of the universe.

Fixed stations on the surface can also probe the deep interior of the Moon. During the Apollo missions, an attempt was made to measure the amount of heat flow in the crust. This heat is produced mostly by the very slow decay of radioactive elements within the Moon. Precise measurements of heat flow allow us to estimate the amounts of these heat-producing elements, which are important constraints to lunar origin. The two measurements of heat flow made on Apollo (near the rim of the Imbrium basin on *Apollo 15* and just inside the Serenitatis basin on *Apollo 17*) were both near mare-highland boundaries, and it has been suggested that a "focusing" effect at such a margin might artificially enhance heat flow here to greater amounts than represen-tative of values in the Moon as a whole. It is important to get a precise, averaged value of the global heat flow to constrain the Moon's bulk composition.

"Shake and Bake": The Deep Interior of the Moon from a Surface Network

If two or more fixed stations make coordinated or linked mea-surements, they constitute a surface network. The first surface network on the Moon was the system of ALSEP stations set up and left on the Moon by the Apollo astronauts (Fig. 3.3). This network of five stations gave us our first measurements of the seismicity of the Moon, its heat flow, the magnetic fields local to the sites, and the tenuous lunar atmosphere. The Apollo ALSEP network was deployed across a small portion of the near side and was operated for six years. To understand the interior and other whole-body properties more completely, we need a global, long-lived surface network.

On Earth, study of seismic events (earthquakes) provides us with our best picture of the structure of the deep interior, includ-ing the primary subdivision of Earth into crust, mantle, and core. The speed at which seismic waves travel through Earth can be measured by the times that they take to travel known distances. This speed depends on the density of the rocks in the interior. Because different rock types have different known densi-ties, we infer the compositions of the mantle and the core from the travel times of seismic waves. In addition, certain seismic waves ("shear" waves or s-waves) cannot travel through liquid, and it is from this fact that we know that the outer core of Earth is made up of *liquid* iron-nickel metal.

We learned from the Apollo ALSEP network that the Moon is extremely quiet in terms of seismic activity. Although "moon-quakes" are rare, they do occur. Moreover, the constant bom-bardment of the Moon by meteoroids provides a source of seis-mic waves that we can use to "see" the crust and mantle (we are not certain if the Moon has a metal core, see Chapter 7). We also *created* seismic events by crashing the used Saturn rockets and Lunar Modules into the Moon, creating a source of known seismic energy and allowing us to determine the thickness of the crust and the approximate depth of shattering from the heavy bombardment (see Chapter 6). A single, natural impact on the far side in July 1972 told us that the Moon may have a small (less than 400 km in radius) metal core and that this core may be partly molten. However, because this interpretation is based on only one event, we cannot be sure that it is correct.

A new global network will reveal the interior structure of the Moon in unprecedented detail. On the basis of gravity mapping from orbit, we suspect that the crustal thickness varies by a wide amount, from as little as a few tens of kilometers under the mare basins to over 100 km in the high portions of the far side. A global seismic net will allow us to more directly determine the true variation in crustal thickness—an important data set for reconstructing the history and internal workings of any planet. It will enlighten us on the character of the magma ocean and on the excavation of large impacts. We can answer once and for all the question of the existence of a metal or sulfide core. The manner in which seismic waves are transmitted through the core tells us whether it is molten or partly molten. The detailed structure of the mantle contains many clues to its own evolution, including the identification of zones from which mare basalts are derived. All of these features can be resolved from a seismic network.

Study of the lunar atmosphere is another task well suited to a global network. The extremely tenuous gases that make up the atmosphere may vary in composition and concentration in space and in time. Continuous monitoring by stations all over the Moon will allow us to understand where the atmosphere comes from, how it behaves, and how portions of it are lost with time. We should characterize the lunar atmosphere before conducting a large-scale, human return to the Moon because the gases emitted by landing rockets and the outpost itself will probably mask the nature of the pristine atmosphere. The interaction of the Moon with the long, weak "tail" of the magnetic field of Earth (Fig. 10.3) may also be studied by a global network. This feature makes up part of the magnetosphere that we will monitor and observe from the Moon.

We need to know the heat flow of the Moon to better estimate its bulk composition. To get the best possible estimate of the global heat flow and to assure ourselves that we are not measuring some anomalous area, we should conduct measurements of heat flow at many different sites, all over the Moon. As the global network is emplaced, heat flow probes should be part of each of the station packages.

Constructing this global network will be a long-term project, taking many years. Emplacing the ALSEP stations involved a considerable amount of effort on the part of the Apollo astronauts. The sensitive instruments needed to make these measurements require gentle treatment and careful alignment during installation. It is not certain that this installation can be accomplished with robotic missions alone, but we may be able to get a rudimentary network going by erecting a stripped-down version with automated spacecraft. However, the long-term measurements that will address fundamental questions in lunar science require delicate, precision instruments. Perhaps robotic teleoperations will permit the "human-like" installation of these advanced stations without the need to send people to many different sites all over the Moon (a desirable capability for exploration in any event). The establishment of this network is a task essential for the long-term exploration of the Moon.

Surface Rovers: Traversing the Moon

The Moon is varied and complex on many different scales. When we study rock types or processes, we must assure ourselves that the inferences that we make have applicability to much wider areas than a hand sample or a microscope slide. One way to accomplish such assurance is to roam the surface of the Moon, making observations and measurements at many different sites. Then we can assess the continuity of rock units and make sure that an ejecta blanket from an impact crater, or a lava flow, has been adequately sampled, studied, and comprehended.

Rovers could be potentially any size and be specialized for different missions (Fig. 10.4). For exploration, we need mobility that allows us to travel many kilometers. We must understand how the Moon varies on local, regional, and global levels. Rovers are particularly good for the local to regional scales of character-

Figure 10.4. A lander and a surface rover. Small surface rovers can accomplish a lot of exploration at minimal risk and cost. We can equip such rovers with instruments to measure surface properties such as composition and physical nature. By conducting initial reconnaissance with robotic rovers, we can ensure that subsequent exploration by people would be much more productive.

ization, whereas orbiters and networks characterize well the regional to global scales of variation. A properly equipped rover can measure the composition and physical properties of many sites. Such traversing can be largely or partly automated, or the rovers can be under complete human remote control (teleoperation). We can make specific measurements, testing ideas about how a certain site developed, or carry out general reconnaissance, making the same set of measurements at each site. Surface roving can be *directed*, as in the surveying of a future exploration site or resource-extraction prospect, or *exploratory*, as in wandering down the interesting "back alleys" of the Moon, discovering new wonders.

A variety of instruments on a rover can produce many different kinds of information. We can measure the chemistry of surface soils and rocks with X-ray and gamma-ray instruments (Fig. 8.2). Spectral measurements can tell us about the mineral makeup of deposits. We can use these instruments together to assess the composition of rock units over wide areas. Rovers can also carry instruments to measure the amounts of solar wind gas and other volatile elements in the soil—information that is important for locating areas to extract hydrogen. The amounts and the fate of volatile elements, such as those associated with the dark mantle deposits, are important scientific questions. A rover mission to such a deposit could measure the contents of volatile elements and determine how they vary laterally and vertically. Such data would yield insight into the processes associated with ash eruption on the Moon.

A rover could also probe the subsurface of the Moon. A technique commonly used to image the subsurface on Earth is *seismic profiling* (also called *active seismometry*). In this process a line or array of geophones, or small seismometers, is laid out in a line or grid on the surface (Fig. 10.5). Explosive charges are then detonated to produce small "moonquakes" of precisely known energy. The seismic waves from these charges travel through the subsurface and are "heard" by the array of geophones. From the travel times and scattering properties of the seismic waves, we can determine the subsurface structure and composition. During the *Apollo 14* and *16* missions, we used this technique on the Moon to characterize the subsurface of those landing sites. Long-distance rover traverses could extend these results to regional scales.

Surface rovers can conduct surveys at sites of high scientific interest before people are sent there. The subsequent human explorations will then be much more profitable. For complex areas this "prereconnaissance" can direct the proper sequence of exploration by determining which investigations are likely to achieve the most results and in which order they should be done. By certifying the sites of resource extraction, we can save time and effort by ensuring that only the best prospects are mined. Investigating the physical properties of habitat sites with rovers will

Figure 10.5. An astronaut laying out a seismic line to record the moonquakes that will be created by the detonation of explosive charges. This technique, used by geophysicists, is the "transect," in which a line on a planet's surface is studied and profiled. Study of the resulting waves can delineate the subsurface geology.

reduce the danger caused by unsuspected hazards. Finally, the inevitable desire to roam free over immense regions of unknown territory (pure exploration) will be partly satisfied through the use of robotic rovers to travel and characterize distant sites all over the Moon.

Sample Returns: The Need for More Pieces of the Moon

Geologists collect samples because we cannot drag into the field the heavy and complex equipment needed to make modern analyses. We examine rocks and soils at all scales, in some cases taking pieces of the Moon apart atom by atom. A collection of samples provides a permanent reference set that can be studied by all of the analytical techniques known to man and in the future by techniques yet to be invented. We still study the samples returned by the Apollo missions, 25 years after they were collected. New techniques are continually applied to the samples, and we are still discovering new facts about the Moon and its evolution. A sample collection provides us with the means to test predictions of models of evolution because it is always available, convenient, and ready for detailed examination and experimentation.

Samples can be collected by a variety of different techniques. Having people pick up rocks is an obvious way to collect samples, as was shown during the Apollo missions. But having robotic spacecraft return drill cores of the regolith was also shown to be an effective means of sample collection by the Soviet Luna missions. Drilling is not required for most investigations. A spacecraft that can land within a designated area and "grab" and return a few handfuls of soil (Fig. 10.6) easily satisfies most scientific purposes. Robotic sample returns can allow us to characterize a wide variety of sites from many different types of terrain.

Certain sampling targets on the Moon are more appropriate for robotic missions than they are for human missions because of their comparative simplicity. Because we do not know the absolute age of the youngest lavas, an extremely important target for sample return is the youngest lava flow on the Moon. Such a flow could be identified by looking for the mare area that has the lowest density of impact craters. A candidate site is near the crater Lichtenberg, where the mare flow covers the ejecta of, and is therefore younger than, a rayed crater (Fig. 5.12). Because the regolith is made from the mare lava bedrock (along with lesser amounts of the underlying rock and a few exotic fragments), a simple robotic mission could get pieces of this lava flow for age determination in the laboratory. Thus the robot probe is completely adequate for achieving the principal objectives of such a mission. Sending a human mission to such a geologically simple site would be an inefficient use of its superior capability.

Combining a suitably equipped rover with a sample-return mission will produce a highly capable tool for exploration. A rover's instruments can sense a variety of chemical and mineral properties and thus will ensure that all the representative and characteristic rocks, and also exotic or rare materials, are sampled and that we do not return many kilograms of the same

Figure 10.6. A robotic sample returner. Small spacecraft similar in concept to the Soviet Luna sample returners (Fig. 3.11) could be used to collect "grab samples" of regional units of simple geological context. Such a mission element would greatly enhance our ability to decipher the Moon's complex history.

material. By taking the minimum amount of sample needed to characterize each geological unit, we will reduce the total mass of samples returned to Earth. For geological characterization, a reconnaissance sample might be as small as a couple of kilograms of soil and a few walnut-sized rocks.

More extensive or complex units require more sample mass, probably in the form of large rock samples for complex, highland breccias. In general, a small quantity of a wide variety of material is preferable to large amounts of the same thing. For resource evaluation or engineering experiments, sample requirements are different. Prospecting requires a large amount of sample to enable experiments of resource extraction at scales large enough to give interpretable results. Similarly, certain engineering experiments, such as those conducted to determine the physical properties of rock and soil, could require many kilograms of sampled material.

Samples can be collected in different ways. The "grab" sample, or simple scoop of regolith material, serves nearly all scientific reconnaissance and engineering requirements (Fig. 10.6). Only with more advanced geological exploration are multiple rock samples or regolith drill cores required. It is also important to document *exactly* where samples have been collected because context can be everything in geology. Typically, documentation can be done by carefully imaging the area sampled before and after the sample is collected. Any supplementary or supporting data, such as compositional measurements, should be tagged so that the sample can be reconstructed into its original setting at a later time.

The collection of samples is an ongoing requirement of lunar exploration. Even as we develop more sophisticated sensors and techniques to make detailed and precise measurements in the field, we should always collect samples, if for no other reason than to ensure that we will not have to repeat fieldwork or re-visit sites (unless we want to). Sometimes it is more convenient to go to the sample vault and make the one measurement that we thought irrelevant or unnecessary than to have to plan a return mission or visit to a distant site. We are never smart enough to anticipate all of the questions, let alone the answers, for a given site. Collecting samples allows us to reconsider our explorations and to make sure that we are always solving the problems that we *think* we are solving.

Human Exploration of the Moon

It is astonishing to me that, more than 30 years after the first human flight in Earth orbit, the debate still rages as to whether people have any role in the exploration of space. I believe that this issue is still debated for two reasons. First, the full potential for human exploration has not been exploited in the manned space program to date. Many (though not all) of the tasks our astronauts do in space could be accomplished by robotic means. This state of affairs leads critics of human spaceflight to argue that there is no significant role for people in space exploration. Second, many of the most persistent critics of manned space-flight have both vested interests and ulterior motives. They suppose that if the human space program were to be canceled, they would have abundant resources for their own projects, whether in space or on Earth.

Just what unique capabilities *do* people bring to the tasks of exploration? This question is meaningless outside the context of the objectives of a mission. People can actually get in the way of certain measurements or experiments. If we want to study the lunar atmosphere, the gases emitted by a space-suited astronaut would completely swamp the atmosphere, at least locally. Likewise, astronauts cannot directly sense the plasmas of interplanetary space or magnetic fields; they can only deploy instruments designed to measure these properties.

Comprehending the three-dimensional makeup of the crust and reconstructing the history and evolution of the Moon are tasks uniquely suited to human explorers. The study of rocks in their natural setting is called *field geology*. Contrary to the beliefs of many (including some scientists!), fieldwork is *not* the same as sample collection. The latter merely refers to picking up rocks and soil. Field study includes that activity, but to a greater extent it is an attempt to understand *completely* the site in question, a goal that includes the study of the original orientations of rock bodies, the contacts between different units, the physical properties of a site, and a visualization of how a site fits into a regional context. Field study is protracted and involves many visits and revisits to selected sites. It could require many complicated techniques and different measurements at a single site or the same, simple measurement at many different sites. It is an ongoing process because the questions raised as a result of our studies require continual reevaluation in light of new results.

Fieldwork is intensely interactive. Small tasks are performed, mentally evaluated, and reperformed. The task list is constantly changed and revised on the basis of immediate sensory input but also according to the changing and evolving conceptual framework. In short, fieldwork is an activity that takes advantage of the unique ability of people—the ability to *think*. We have yet to build a robot that has anywhere near the capability of people to reason, make judgments based on experience, or bring specific technical expertise to bear on surface exploration. The great promise of machine intelligence remains just that—a promise.

In a few instances in the past, people have been turned loose to exercise their creativity in space. Astronauts on the Apollo missions tried to make the most of their limited time on the Moon, cramming observations into their crowded, minute-by-minute schedule (Fig. 10.7). If we want to be able to take advantage of human capabilities in space, we must give people a much looser rein. Fieldwork is a partly directed but flexible activity. Much of the time involved in the field is spent looking and thinking. I contend that such effort, which many engineers and program managers see as "wasting time," is the true essence of field science and is mandatory for the making of significant, exciting discoveries.

Complex areas are well suited to human exploration. For example, a crater ejecta blanket is made up of a myriad of complex, altered, shattered, and melted rocks. Many different types of rock can be present, and they must be recognized and cataloged. Representative and exotic samples are collected. The vertical and lateral variations in the rock unit must be comprehended. Such comprehension can be achieved by traversing (by foot or vehicle) in different directions (Fig. 10.8) to understand the scales and types of variation. Subtle changes in color, texture, and composition are noted. A mental image (sometimes of great complexity) is built up and is constantly revised, tested, and updated. Sometimes a great revelation occurs in the field, resulting in conceptual breakthroughs, but it is more likely that patient and careful work over a long time yields the most insight. In the field the senses are assaulted by an overwhelming mass of data and input. It takes a *human* mind to separate the important from the trivial—to pick out the single, key piece of the puzzle from the myriad irrelevant facts that could be collected or measured.

The knowledge and the judgment that people bring to exploration are essential for us to fully comprehend the Moon. But does this mean that people need be *physically* present? The concept of *telepresence*, in which a suitably equipped robot acts as a human surrogate, under the complete control of a person at a remote location (Fig. 9.9), has been advocated as a technique we should use to explore the planets. Telepresence has many advantages. Robots can be physically strong, possess advanced sensory capabilities, and be made impervious to the environmental extremes

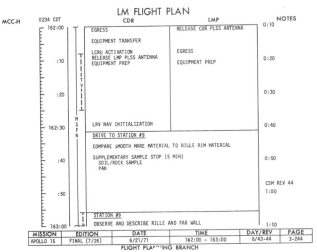

LM FLIGHT PLAN

```
MCC-H   0334 CDT                    CDR                    LMP              NOTES
        163:00                                                             1:10
              | | |  500MM PHOTOGRAPHY
              | | |  COMPREHENSIVE SAMPLE
              | | |  SINGLE (DOUBLE) CORE TUBE
              | | |  PAN
              | | |  DOCUMENTED SAMPLING OF CRATER AT EDGE OF RILLE
          :10 | | |  PENETROMETER                                          1:20
              | | |  POSSIBLE PAN OF EDGE OF CRATER
```

found on the Moon. The data returned from the robot could be made available to many people on Earth, allowing them to share in the experience of the exploration of another world. Telepresence might be the ultimate resolution of the man-machine debate, melding together the advantages of robots (strength, endurance, and sensory powers) and those of humans (imagination, reasoning, and the application of expert knowledge to the solution of problems).

Telepresence may be an important tool for the exploration of planetary surfaces, but many unknowns remain. Any significant separation between robot and operator causes a time delay and could seriously degrade the telepresence effect. We do not fully understand how people who do field study construct their mental frameworks. Is the process entirely mental, or does it result from the intimate interaction of people with their environment? What is the maximum amount of time delay that can be tolerated? We need to conduct experiments to understand these effects before we count on telepresence as the principal tool for surface exploration.

Another advantage that people bring to space exploration is their ability to fix and adjust things (Fig. 10.2). Sometimes damage to equipment is beyond repair, at least in space, but the history of manned spaceflight is strewn with examples of astronauts administering a "swift kick" to get a balky piece of equipment running. The nature of field repair brings out qualities that people possess: fine-scale dexterity, judgment, observation and the recognition of key clues, and the selective application of knowledge to solve problems. People are good at innovation, and astronauts are sometimes able to fix equipment at the scene. During the *Apollo 17* mission in 1972, the astronauts fixed a broken fender on the wheel of the lunar rover by improvising a

Figure 10.7 (opposite). An astronaut's checklist. During the Apollo explorations, every minute of time spent on the Moon was carefully budgeted. The astronauts carried detailed checklists on their wrists, telling them their assigned tasks at each moment. Such thorough planning, though appropriate for the limited-duration Apollo missions, is *not* desirable for surface exploration. People must be free to make significant discoveries when we return to the Moon.

Figure 10.8. Traversing the Moon. Long-range explorations by people are required to fully explore the Moon. Surface traversing in pressurized rovers is the best way to characterize the regional geological structure.

new replacement fender out of old maps. Field repairs often involve solutions to problems unanticipated by mission planners back on Earth.

People in space are important for another reason. The astronauts in orbit or on the Moon are our surrogates, our human window onto the universe. Machines can collect data that, in turn, can be processed into fantastic, stereo, Technicolor views of the solar system. We can thus uncover subtle facts about the Moon, sample the atmosphere, image the interior with seismic waves, and uncover many hidden secrets. But our robotic spacecraft cannot inspire or convey the same impact or impression that people can. The exploration impulse is not a totally rational one, and the power of inspiration is largely emotional. As someone once said: "There are no ticker-tape parades for robots." Unlike many of my scientific colleagues, I do not despise the irrational impulses that drive us to "go there ourselves." Such an impulse has served humanity well during our struggle to conquer the globe. As we contemplate extending this struggle into the solar system, by going to the Moon, we are continuing a great tradition: once again to learn and to seek the unknown beyond our boundaries. A man's reach should exceed his grasp, or what's a heaven for?

The Utilization, Industrialization, and Colonization of the Moon

Extracting products from the Moon to sustain life and operate in space will be one of the principal tasks of a lunar return. The early use of lunar resources is likely to be limited to materials that require little processing or handling. An example is the use of bulk soil to shield the surface habitat. I suggest that a general strategy for using lunar resources should be to do the easy things first, gain experience, and expand operations as abilities and needs develop.

The production of oxygen for rocket fuel and life support is likely to be the first large-scale operation to be undertaken on the Moon, not only because such products are urgently needed but also because we know enough about the extraction technology (Fig. 9.6) to have a reasonable assurance of success. Selection of the appropriate method of extraction should depend on the outcome of carefully structured experiments. We need operational experience to learn which process is the easiest and most efficient. Oxygen produced on the Moon would greatly facilitate access to the Moon and would permit a giant leap in the capability of the base, both in terms of the mass needed from Earth to support surface operations and in terms of the number of people an outpost could support.

The next most likely activities would be making building materials (ceramics and bricks) for permanent establishments and installations and extracting solar wind hydrogen from the soil. The production of building material is a logical outgrowth of using regolith for shielding. Blocks could first be used to make radiation shelters and observatory walls. The construction of a network of surface roads linking various outposts and installations would be a high priority; roads could be made simply by bulldozing, grading, and flash-melting the surface soil into a glassy layer. Shelter construction might follow patterns used in the making of adobe structures on Earth. As lunar materials are used more widely, we will probably devise innovations and processes that we cannot now envision.

The problems associated with the production of hydrogen are broadly understood. The very low concentration of hydrogen in the lunar soil requires a large-scale operation for the production of significant amounts of this material. Because of this investment in time and resource, it is prudent to include consideration of the use of the by-products from hydrogen mining. Sulfur is a major component of the soil at these levels of processing; this element has many industrial uses and could even be used as a rocket propellant for short-distance "hops" across the face of the Moon. The nitrogen and carbon extracted along with the hydrogen should be carefully recovered because both are essential in the production of food and other materials, such as plastic, for the lunar base.

The first production of energy on the lunar surface will probably be geared toward its use on the Moon. Solar thermal and solar electric power are likely to be the principal early uses of the energy resources of the Moon (Fig. 9.7). Because of the 14-day lunar night, some method will be needed to allow the base to live through the period of darkness, and fuel cells or nuclear reactors are likely to be required for the initial installation. An exception to this requirement exists at the south pole, where a crater rim at the pole may be in more or less continuous sunlight. At this locality a base could use electrical power produced exclusively from sunlight.

After experimentation in the manufacture of solar electric cells on the Moon, we may consider constructing large solar arrays in place on the surface. Initially, such arrays would supply electrical power to the base and its surrounding installations. If the construction of large arrays appears to be feasible, we could consider the industrial-scale production of electrical power on the Moon. Again, such power production should be phased in gradually, after we gain experience in working in the unfamiliar and exotic lunar environment. As large-scale hydrogen extraction becomes well established, we should carefully separate ^{3}He from the collected gases of the regolith. If markets develop for the terrestrial use of ^{3}He, we will be in a position to produce ^{3}He on the Moon by taking advantage of the technology base we will have built for hydrogen mining.

This phased approach to resource utilization avoids massive investment in processes or techniques that may later turn out to be less than desirable. By using the appropriate materials and energies at the appropriate times, we learn how to live and work in space. We increase the capability of the lunar installation gradually, using the indigenous resources to leverage our investment while simultaneously learning how to make money in space. In other words, the tail shouldn't wag the dog; we should use lunar resources to help the base develop and flourish and *then* produce for export and profit. We should not go to the Moon to mine it at the outset.

The idea that we can live, inhabit, and use the Moon to create a space-faring civilization is offensive to some people. Over the last few years a lunar-preservation movement (we'll call its members "grays," after the "green" environmental movement here on Earth) has emerged that opposes the industrial development of the Moon. The claim of the "grays" is that we have no right to colonize, mine, or change the Moon and other planets. They argue from a moral, not a scientific or technical, perspective. The readers of this book will have to judge for themselves whether these people have a point or not. I note only that, inevitably, where man has seen, man has gone. And where man has gone, man has changed. Eventually, someone will do the things described in this section. The process may be as chaotic and dynamic as the European conquest of the New World, or it may be as orderly, regulated, and bureaucratic as the joint investment of Antarctica by the countries of the world. Only time will tell.

 Chapter 11

When?
What's Holding Us Back?

Having looked at the why and the how of a return to the Moon, we now ask the inevitable next question: When? The "lunar underground" has been trying, to no avail, to get a return to the Moon started ever since the end of the Apollo program. President George Bush tried in 1989; this effort went nowhere. The Clementine mission succeeded in carrying out the first global, remote-sensing reconnaissance of the Moon, in 1994, but an imagined follow-up to Clementine has yet to materialize. The Japanese apparently are very interested in going to the Moon, with a small, hard-landing penetrator mission scheduled for 1997. There has been discussion about a European lunar program, but the first mission, a remote-sensing orbiter, is not scheduled to fly until after the turn of the century. Once again, all of this is merely talk.

Is the vision of lunar bases, self-sufficiency in space, and planetary colonization all a pipe dream? Are the advocates of a return to the Moon deluding themselves? Much of the anticipation of a lunar return draws on an analogy to the Apollo program, in which a dynamic president challenged a future-oriented, brash, and confident country to go to the Moon and to do it within a decade. We are not that same optimistic nation. A similar challenge by another president to construct a space station within a decade, a task that is technically much *easier* than going to the Moon, never even got close to its deadline and, at this writing (early 1996), has yet to see the launch of a single element.

So what *is* the problem? We need to understand the answer to this question before we can know when we are likely to return to the Moon. Expense is one obstacle, but this difficulty is a short-term one and is addressable by a number of different means. Other, more problematical roadblocks involve the nature of our society. Let's examine the obstacles in turn.

Means and Ends

It is very expensive to go into space. The Space Shuttle can send payloads of up to 25 tons, at an average cost of about $5,000 per pound, to low Earth orbit. Other launch vehicles are not much cheaper. Typical costs for Europe's *Arianne 5* are about $2,000–3,000 per pound, and the Russian Proton still requires almost $1,000 per pound. Economies of scale might be possible. For example, a Russian Energia can put almost 100 tons into Earth orbit for perhaps $600–800 per pound (it is very difficult to estimate these costs for surplus Soviet equipment). It takes about 5 pounds in low Earth orbit (mostly fuel, tanks, and rocket engines) to get 1 pound to the lunar surface. We must carry everything with us (at least at first), including habitats, air, water, food, equipment, and the fuel to get back home. As the list builds, one can see why Moon programs are said to cost billions.

The actual cost of the *energy* or fuel to put a payload on the Moon is only a few dollars. So where do these astronomical (no pun intended) costs come from? Spaceflight is expensive because the current engineering state of the art requires complex machines serviced by a small army of workers for rocket assembly, payload preparation, launch, and flight. This "marching army" works many months to do these tasks, and these people must be paid. We must carefully assemble a highly trained elite of technicians, engineers, and scientists to service and maintain a launch facility. The immense infrastructure of ground support people, office workers, accountants, managers, coordinators, counselors, and the proliferating myriads that make up the support base of a space-launch capability comes with this army.

This infrastructure must be paid, whether or not a rocket is launched. This expense is rolled into the per-pound cost of payload to orbit. So the expense of spaceflight is not simply a function of which launch vehicle is flown or how a fleet of launch vehicles are flown. It is a complex number that reflects a culture—an attitude about how space business is conducted, what its perceived mission is, and how the various necessary, complex tasks of spaceflight capability are managed. This complexity illustrates one of the reasons we cannot precisely determine the per-flight cost of a Space Shuttle mission.

So how can we lower the cost of spaceflight? We need to do a lot of different things. A new launch vehicle that takes advantage of new automation technology to greatly reduce the manpower costs of ground support is *essential*. An example of such a launch vehicle is the *Delta Clipper* (or *DCX*, Fig. 11.1), an experimental prototype of a single-stage to orbit launcher, built by the Department of Defense's Strategic Defense Initiative Organization, the same people who flew the Clementine mission to the Moon. The *Delta Clipper* is designed to launch small payloads on a completely reusable, remote-controlled rocket. Assembly, checkout, and launch of this rocket requires about 50 people (a Space Shuttle launch process requires almost 5,000 people). The *Delta Clipper* takes off and lands vertically, so its spaceport can be located anywhere. The current *Delta Clipper* is designed to test its engines and conduct short-range hops. An orbital version of the *DCX* could be flying within five years.

There are other options to lower launch costs. For very large efforts, such as a return to the Moon, a heavy-lift booster (more than 100 tons to low Earth orbit) could produce significant economies of scale. The Russian Energia is the only existing heavy-lift rocket (80–90 tons to LEO) in the world today, but because it has flown only twice, there are serious questions about its reliability. During the work of the Synthesis Group in 1991 (see Chapter 8), we determined that it is possible in principle to reopen the production lines for Rockwell's F-1 engine (Fig. 11.2), the heart of the first stage of the U.S. *Saturn 5* and an engine around which a new heavy-lift booster could be made. Updating the *Saturn 5* with new materials for fuel tanks and advanced flight electronics technology would produce a launch vehicle that could put 150–200 tons into Earth orbit for perhaps less than a few hundred dollars per pound. Such a vehicle could put us back on the Moon with only two launches within five to seven years of the program's initiation.

The basic challenge in launch services is to move the resources we are now investing in development and performance into pro-

Figure 11.1. The experimental *Delta Clipper (DCX)* launch vehicle, developed by the Strategic Defense Initiative Organization. This vehicle is a prototype of a reusable, single-stage to orbit launcher that, because of automated checkout procedures, requires only a few tens of people to operate, as opposed to the thousands of workers needed for current launch systems. This makes launch costs much less expensive, one of the first requirements for a permanent human presence in space.

Figure 11.2. The F-1 engine. Another way to lower launch costs is to send up very large payloads all at once, as was done for the U.S. Apollo lunar craft, the U.S. Skylab space station, and the Soviet Mir space station. This massive F-1 engine, one of a cluster of five used in the first stage of the *Saturn 5* heavy-lift launch vehicle, developed 1.5 million pounds of thrust. This engine could be produced again to re-create the heavy-lift capability that the United States once had.

duction and operations. We need to open production lines to build simple, relatively cheap rocket parts and then mass-produce them. Another concept for inexpensive launch vehicles uses the simplest and most reliable technology (for example, using rocket engines that burn kerosene, a cheap and relatively safe fuel). These launchers would contain only minimal electronics and guidance and control packages, all of which would be modular and mass-produced for economy. Such launchers, called "Big Dumb Boosters," can be built to carry very large (100 ton) payloads. The emphasis is on modularity, automated assembly and check-out, and frequent launches. As another technique to lower the costs of space access, the Big Dumb Booster should be thoroughly investigated.

Finally, a flexible architecture for lunar return allows us to do it in steps. We can spread out or compress the flight schedule in accordance with available resources. Both approaches have advantages and drawbacks. In general, doing it quickly means doing it cheaply. Clearly, *how* a program is organized and run will have great impact on its cost. If we decide that a return to the Moon is a desirable goal, we can make it much less expensive than many opponents estimate. But money is not the only problem.

Failure of Vision

We live in a society that thrives on and expects instant gratification. Future planning and investment is based on short-term return. Our leaders engage in constant "crisis management." The annual federal budget process consumes the vast bulk of the legislative schedule. Congress inadvertently adds to mission costs when it attempts to manage technical projects financed by federal dollars, thus ensuring that things we used to be able to do (like go to the Moon) are no longer "possible" or "feasible." Our preoccupation with the near term has been harmful, if not disastrous, for our nation and has contributed to our unchecked and increasing national debt. We need to find ways to energize our people and economy with an eye toward fiscal recovery and future prosperity.

The absence of any long-term view of what we should be doing for the future—our *attitude* about the future—is a failure of vision. In the United States of the 1960s, we were confident that the future held great promise, even if we did not know what form that better, brighter future would take. When President John F. Kennedy issued his challenge to go to the Moon and do it within a decade, the nation and Congress were eager to respond. Clearly, this was the right thing to do. The country looked forward to and anticipated the future; we did not dread it.

We now have a different attitude about the future. We worry about and agonize over it. When will we run out of fossil fuels? When will global warming cook us all? When will global cooling freeze us all? How can we compete economically with Japan, Europe, everybody? We do everything about the future except prepare for it. Our leaders no longer articulate bold visions; they try to avert our eyes. They no longer lead; they base their own "principles" on the polls of focus groups.

The space program, more than any other activity undertaken by the federal government, requires a future-oriented vision. One could argue (and many have argued) that the single biggest problem with our space program is the absence of a mission. When NASA was formed almost 40 years ago, it had a mission: beat the Russians, in whatever space arena they chose. A trip to the Moon soon became our focus because it was a challenging goal, just out of reach yet perceived to be reachable. After we won that race, NASA needed a purpose. It quickly pulled together a program of sorts with the Space Shuttle, but repeatedly flying people in low Earth orbit does not a mission make. Unfortunately, the Space Station has the same problem. It makes no sense except in the context of the future movement of humanity out into the solar system.

This failure of vision keeps us doing trivial things in space (and on Earth). We will never return to the Moon, move into the solar system, or strive toward any other worthwhile goal unless the nation as a whole regains its confidence about itself and the future. National goals are set through a well-developed political process, and real leadership from key individuals, such as the president, is extremely important to galvanize public opinion and provide a rallying point for political support. Until the nation has some perceived motivation to go to the Moon, we will not do so.

Suppose that we as a society could agree that a return to the Moon is a desirable goal. Could we undertake such a challenge? Could the United States once again marshal the resources, the expertise, and most important, the *will* to carry out a long-term, focused space goal?

Failure of Nerve

We live in a society that is increasingly averse to, and in many ways has lost its ability to cope with, risk. The intelligent management of risk is essential in any arena, but it is particularly important in dangerous endeavors such as war and exploration. The fear of risk has brought us to near-paralysis in many ways. We are afraid to begin new, large projects because they may end up costing more than we thought. We fear the potential for loss of life, loss of jobs, or loss of status. As a society, we have become risk-averse, a condition that is becoming pathological. This risk-aversion is physical, intellectual, and moral.

We have become physically afraid to do bold things. In January 1986 the tragic *Challenger* explosion was a disaster for the country and the space program. In this accident, the nation's first "teacher in space" was killed, along with six experienced astronauts. Within three months, an unpiloted U.S. Air Force *Titan 4* launch vehicle also exploded, just seconds after liftoff. In the aftermath of these accidents, the nation had *no* capability to launch large payloads to orbit. This paralysis lasted over two years. Space efforts in general and NASA in particular came under intense criticism, and there were calls to terminate the space program. Even after the Space Shuttle started flying again, pundits prophesied doom on some future missions, and programs designed to let private citizens fly in space were suspended indefinitely.

Spaceflight is and always will be hazardous to some degree. Objectively viewed, particularly in the context of the profession of test flight, the Space Shuttle is an extremely reliable and safe launch system and has the lowest failure rate of all the launch systems in the world. Risk management in the context of space travel means the ability to understand as many potential dangers as possible and to minimize and then accept the chances of some disaster. The only perfectly safe mission is the one that never gets launched.

The failure of nerve also involves intellectual timidity. Such fear is often expressed in terms of worry about the costs of space exploration. "We can't afford to go to the Moon!" critics complain. I talk to many different people about space exploration, and I always like to ask them how much they *think* we are spending on the space program, both in total dollars and as a fraction of the federal budget. I find that almost everyone grossly overestimates the amount of spending on the space program. I often hear numbers like 50 percent of the federal budget or hundreds of billions of dollars per year. In fact the budget for NASA (the "civilian" space program) is about $14 billion per year and constitutes less than 1 percent of the federal budget (Fig. 11.3). Expressed as a fraction of the Gross Domestic Product (GDP, the total value of goods and services produced in the country, a measure of our national "wealth"), the NASA budget is much less than one-quarter of one percent ($< 0.25\%$) of the GDP. We could double this amount of spending and not go bankrupt. But throwing money at a problem never solved anything, and doubling the NASA budget would not produce a better or more exciting space program, unless some other changes occur.

The intellectual failure of nerve is not limited to people outside the space program. Setting a long-term goal to provide a "mission" for our space program would do a lot to get us back on track. The current NASA strategic plan (neither "strategic" nor a "plan") lists the five "businesses" of the agency: aeronautics, technology, science, "Mission to Planet Earth," and human spaceflight. Yet listing agency activities does not make a "mission." Arguably, some of these activities could be done more appropriately by other entities. In fact, keeping the agency infrastructure funded and running, regardless of what it is actually *doing*, has become the principal mission of NASA. All of this is intellectual risk-aversion: We don't want (or dare) to say what we really are. (Hint: NASA's "business" should be to explore the universe with people and machines.)

The third aspect of the failure of nerve is a moral failure. Some people contend that, with all of the problems we face on Earth, we should not explore space. They would have all of the

U. S. Federal Budget - "On Budget" expenditures 1993

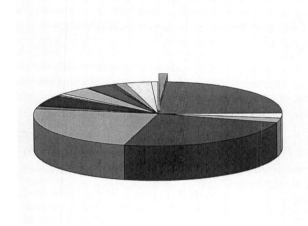

NASA
Health and Human Services
Education
HUD
Interest on debt
Defense
Energy
Commerce
Agriculture
Interior
Justice
Labor
State
Transportation
Treasury
EPA
Govt. ops
GSA
OPM
OFA

U. S. Federal Budget - All Expenditures 1993

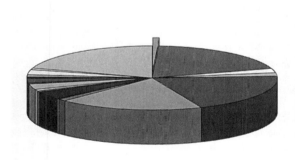

NASA
Health and Human Services
Education
HUD
Interest on debt
Defense
Energy
Commerce
Agriculture
Interior
Justice
Labor
State
Transportation
Treasury
EPA
Govt. ops
GSA
OPM
OFA
Social Security

Figure 11.3. The budget of the National Aeronautics and Space Administration (NASA) as a function of discretionary and total federal spending (FY1994 dollars). We spend less than 1 percent of the federal budget on the space program.

"tremendous resources" that we "pour" into space spent on various social projects. These beliefs assume that space exploration is of intrinsically less societal value than other government activities, and thus this idea constitutes a moral argument. For the people who make such an argument, space exploration is a frivolous activity and spaceflight is something that we should not be doing even if it cost nothing.

It is always difficult to "justify" exploration on cost-benefit grounds, but let me try. We invest in exploration (and basic research, for that matter) primarily for one simple reason: We are not smart enough to know ahead of time all of our needs and wants. Knowledge always pays off, sometime and somehow. Exploration gives us new knowledge, but more important, it *broadens our imagination* so that we can see solutions to problems that we would not have imagined otherwise. Often some of our toughest problems are created because we are not posing the right question. The greater our imagination, the more likely we are to at least recognize the proper question or to pose a question that we can answer or a problem that we can solve. Because space exploration is a challenging endeavor that calls on the best qualities that people have to offer, such as ingenuity and perseverance, we will always be forced to confront and conquer the unknown. Exploration is a very human thing to do!

What's the Solution?

None of the problems outlined in this chapter have easy, simple, or single solutions. Some difficulties are cultural and thus are part of a larger set of attitudes within society as a whole. Others are more easily addressed, and solving them could go a long way toward getting us back on track. Is there a way to reenergize our space effort and give it some direction? What is the path back to the Moon?

Although predicting political landscapes is fraught with pitfalls, I would suggest that a national, government-run effort to return to the Moon, an undertaking similar in pattern to the Apollo program, is unlikely to occur. Apollo came about because of a unique set of political and technical circumstances: We were the right nation in the right place at the right time, with the right spark, to go to the Moon. Because political circumstances are inherently unpredictable, similar conditions may arise again; advocates of a lunar return tend to believe this will happen, and they hope the country will be ready. I doubt that such an opportunity will arise again—and even if it does, I am skeptical that we will respond effectively to it (witness the rise and fall of SEI, Chapter 8).

The history of the United States shows that only national security is a strong motivator for the undertaking *and the continuing support* of large engineering projects. Successful technical projects that we have finished (such as the Panama Canal, the atom bomb, and the Moon landing) were all ultimately a response by this nation to a security threat, real or imagined. The history of large science or engineering projects undertaken outside such a motive is littered with discarded and uncompleted tasks, such as "Mohole" (a plan to drill through Earth's crust) or the Semiconducting Supercollider. For now, there is no national security interest associated with a lunar return, so I believe that we are unlikely to return to the Moon with an effort modeled along the lines of the Apollo program.

Some in the space community claim that the only way we can possibly afford a lunar base or other ambitious space goals is through international cooperation. They contend that by having many different countries share costs, space activities become affordable. However, in many ways, international cooperation in large space efforts costs us *more* money. As a price for their support, each partner demands an equal voice in how the project is run. The result is indecisive consensus management and increased bureaucracy. Programs require increased time and more personnel. There is duplication in capabilities and infrastructure. By the time flight hardware is actually launched and operating, it is inevitably more costly and less capable than originally planned.

International cooperation in space can be successful, on a smaller scale. Scientific participation in the Apollo program was international in scope, with foreign investigators planning surface experiments and analyzing returned samples. In a more recent example, the global mapping of the Moon by the Clementine mission was largely possible because the spacecraft flew with a data-compression microchip provided by the French Space Agency. A productive international partnership is possible, but spaceflight is largely an engineering task, and such work is not amenable to consensus management. At the project level, one nation or entity should retain decision authority.

Another problem we face in revitalizing our space program is that the capabilities of spaceflight are too centralized. In this country, NASA has the sole authority to conduct human and robotic space-exploration missions. In other words, it has a monopoly. Monopolies possess many enduring characteristics. They tend to have fixed ways of doing business and are resistant to change. We need to find some way of widening the field of play and permitting many other entities, both public and private, to enter the space-exploration game. Competition creates both drive and excitement; a little competition would do the space business a world of good.

If I could do one thing that would ensure a return to the Moon, it would be to lower the cost of access to space. By this I do not mean creating a new multibillion-dollar program for a new heavy-lift booster program, and I do not mean cutting costs by factors of a half or a quarter. We must lower launch costs by factors of 50 to 100. Such a drastic reduction of launch costs will permit a stagnant commercial launch industry to explode with activity. Once the cost is $50 instead of $5,000 to launch a pound into Earth orbit, many things become possible that now appear to be science fiction. Anyone would be able to scrape together enough money to get significant payloads into orbit, and the private sector would begin to move industry into space. Once started, the profit motive would largely finance a genuine movement of humanity into the solar system, of which the Moon is the logical first way station.

In short, I am arguing for a return to the Moon based on a model that has more in common with our nation's westward movement than with a massive government research-and-development effort. During the western migration the government, both military and civilian, played early, important roles in conducting the initial explorations and in opening the frontier, and it then held continuing roles in maintaining the peace. But it was the common multitudes—the settlers, builders, entrepreneurs, and risk-takers—who tamed the wilderness and created our nation's wealth. What is holding us back in space is the lack of a Conestoga wagon, in this case, cheap and available access to orbit. We need a "Volksrocket," a booster that can get privately financed payloads into space cheaply and effectively. I believe that the road back to the Moon does not travel through the Space Shuttle and the Space Station but through the *Delta Clipper*, the Big Dumb Booster, and other launch vehicle alternatives. If many different players are doing many different things in space, the odds are better than even money that there will be breakthroughs. A centrally planned, carefully choreographed effort is more likely to lead down blind alleys.

What should be the role of government in space? I believe it is to do those things that are technically difficult or beyond the ability of the private sector to address. Originally, NASA was a research-and-development agency that invented many of the technologies needed for spaceflight. A government agency could still lead the way in developing advanced propulsion technology, automated systems, planetary surface operations, and a host of other fields. It should *not* be a launch service, a collection agency for remote-sensing data, or a jobs program for technical managers. Clementine was an example of a mission appropriate for government: the transfer of advanced technology (sometimes classified) from the government to everyone. It was a risky mission in that many different pieces of flight hardware, software, and procedures were being used for the first time. Now they have been "qualified" and can be used for a variety of different purposes and missions.

We should return to the Moon in a careful series of small steps taken by both government and industry. *But we have to get started!* The Clementine mission blazed a new trail to the Moon, and we should capitalize on its success. The most appropriate next steps are to finish the global mapping begun by Clementine and then to conduct a series of small, robotic missions to the lunar surface to explore, investigate, and characterize selected areas. A surface mission to the south pole of the Moon could examine the polar deposits and characterize the environment, particularly the areas of permanent darkness and near-permanent sunlight. A properly equipped rover mission to high-titanium maria could measure the contents of solar wind gases in the regolith. Small automated telescopes, remotely operated from Earth, could assess the value of conducting astronomy from the Moon. All of these small missions, combined into a single program, would cost much less than a single Space Shuttle mission. Combined, they would constitute the first coordinated, coherent steps in a journey back to the Moon.

The Moon, the solar system, and the universe beyond are a source of knowledge and wealth. We are fortunate in having a small planet as fascinating and as useful as our Moon right in our own space "backyard." We should return to the Moon—to study it, to use it, to inhabit it. A lunar return can reinvigorate not only our space efforts but also our society. The United States has always been a frontier society. With a return to the Moon, we can once again dream of a frontier, of a horizon beyond which lie unexplored territories and unknown treasures offering limitless opportunities.

Appendix 1

Basic Data about the Moon

Mass	7.35 × 10²² kg (1% mass of Earth)
Radius	1,738 km (27% radius of Earth)
Surface area	3.79 × 10⁷ km² (7% area of Earth)
Density	3,340 kg/m³ (3.34 g/cm³)
Gravity	1.62 m/sec² (0.17 gravity of Earth)
Escape velocity	2.38 km/sec
Orbital velocity	1.68 km/sec
Inclination of spin axis (to Sun)	1.6°
Inclination of orbital plane (to Sun)	5.9°
Distance from Earth	
Closest	356,410 km
Farthest	406,697 km
Orbital eccentricity	0.055
Albedo (fraction light reflected)	0.07–0.24 (average terrae: 0.11– 0.18; average maria: 0.07–0.10)
Rotation period (noon-to-noon; average)	29.53 Earth days (709 hours)
Revolution period (around Earth)	27.3 Earth days (656 hours)
Average surface temperature	107°C (day); −153°C (night)
Surface temperature in polar areas	−30° to −50°C (in light); −230°C (in shadows)

 Appendix 2

Robotic Missions to the Moon

Mission	Launch Date (month/year)	Country	Type
Luna 1	01/59	USSR	flyby
Luna 2	09/59	USSR	hard lander
Luna 3	10/59	USSR	flyby (pictures of far side)
Ranger 3	01/62	USA	hard lander (missed the Moon)
Ranger 4	04/62	USA	hard lander (hit far side)
Ranger 5	10/62	USA	hard lander (missed the Moon)
Luna 4	04/63	USSR	flyby (missed the Moon)
Ranger 6	01/64	USA	hard lander (TV failed)
Ranger 7	07/64	USA	hard lander
Ranger 8	02/65	USA	hard lander
Ranger 9	03/65	USA	hard lander
Luna 5	05/65	USSR	soft lander (crashed)
Luna 6	06/65	USSR	soft lander (missed the Moon)
Zond 3	07/65	USSR	flyby (pictures of far side)
Luna 7	10/65	USSR	soft lander (crashed)
Luna 8	12/65	USSR	soft lander (crashed)
Luna 9	01/66	USSR	first soft landing
Luna 10	03/66	USSR	first lunar orbiter

Mission	Launch Date (month/year)	Country	Type
Surveyor 1	05/66	USA	first American soft lander
Lunar Orbiter 1	08/66	USA	orbiter
Luna 11	08/66	USSR	orbiter
Surveyor 2	09/66	USA	lander (crashed)
Luna 12	10/66	USSR	orbiter
Lunar Orbiter 2	11/66	USA	orbiter
Luna 13	12/66	USSR	soft lander
Lunar Orbiter 3	02/67	USA	orbiter
Surveyor 3	04/67	USA	soft lander
Lunar Orbiter 4	05/67	USA	orbiter
Surveyor 4	07/67	USA	lander (crashed)
Explorer 35	07/67	USA	orbiter
Lunar Orbiter 5	08/67	USA	orbiter
Surveyor 5	09/67	USA	soft lander
Surveyor 6	11/67	USA	soft lander
Surveyor 7	01/68	USA	soft lander
Luna 14	04/68	USSR	orbiter
Zond 5	09/68	USSR	flyby and return
Zond 6	11/68	USSR	flyby and return
Luna 15	07/69	USSR	sample returner (crashed)
Zond 7	08/69	USSR	flyby and return
Luna 16	09/70	USSR	first robotic sample return
Zond 8	10/70	USSR	flyby and return
Luna 17	11/70	USSR	surface rover
Luna 18	09/71	USSR	orbiter(?) (crashed)
Luna 19	09/71	USSR	orbiter
Luna 20	02/72	USSR	sample returner
Luna 21	01/73	USSR	surface rover
Luna 22	05/74	USSR	orbiter
Luna 23	10/74	USSR	sample returner (failed)
Luna 24	08/76	USSR	sample returner
Muses A	01/90	Japan	orbiter
Clementine	01/94	USA	orbiter (global mapping)

Human Missions to the Moon

Mission	Launch Date	Type	Site	Remarks
Apollo 8	12/68	Orbiter	—	First humans to orbit the Moon
Apollo 10	05/69	Orbiter	—	Test of LM in lunar orbit
Apollo 11	07/69	Landing	Mare Tranquillitatis	First manned lunar landing
Apollo 12	11/69	Landing	Oceanus Procellarum	Pinpoint landing near Surveyor 3 spacecraft
Apollo 13	04/70	Flyby	—	Aborted after spacecraft explosion
Apollo 14	01/71	Landing	Fra Mauro	First highlands mission
Apollo 15	07/71	Landing	Hadley-Apennines	First use of rover, extended LM
Apollo 16	04/72	Landing	Descartes	Mission to lunar highlands
Apollo 17	12/72	Landing	Taurus-Littrow	Last Apollo landing; first geologist on the Moon

Note: All missions were launched by the United States.

Appendix 4

Conversion of Units
Metric to English

Until the United States adopts the metric system, we'll have to struggle along with dual systems of measurement. I have tried to use metric units consistently throughout the text, except where common usage demands English units.

Property	Metric Unit	English Unit	Factor
Mass	kilogram (kg)	pound (lb)	1 kg = 2.2046 lbs
Mass	gram (g)	pound (lb)	1 g = 0.0022 lbs
Mass	ton (mt)*	pound (lb)	1 mt = 2,204.6 lbs
Distance	kilometer (km)	mile (mi)	1 km = 0.6214 mi
Length	centimeter (cm)	inch (in)	1 cm = 0.3937 in
Length	millimeter (mm)	inch (in)	1 mm = 0.03937 in
Length	meter (m)	foot (ft)	1 m = 3 28 ft
Length	meter (m)	inch (in)	1 m = 39.37 in
Length	micrometer (λm)	—	1 m = 1$0^{-6}$ m
Temperature**	degree Celsius (°C)	degree Fahrenheit (°F)	T(C) = 5/9 [T(F)−32]

*1,000 kg is known as the metric ton; the English ton is 2,000 lbs.
**On the Celsius scale, water freezes at 0°C (32°F) and boils at 100°C (212°F).

 Appendix 5

Moon Places
Museums, Visitor's Centers, and Worthwhile Places to Visit

I've been to many—but not all—space museums, NASA visitor's centers, and other places that might interest the readers of this book. Here I offer an assessment of their strengths and weaknesses.

Alabama Space and Rocket Center, Marshall Space Flight Center, Huntsville, Alabama

One of the best space museum and visitor's centers in the country, surpassed only by the National Air and Space Museum. The museum is superb; the rocket park has *Saturn 5* plus many other rockets from space history. A bus tour takes you through Marshall Space Flight Center, where Wernher von Braun and his team built the Saturn rocket. The U.S. Space Camp is located here.

Craters of the Moon National Monument, near Arco, Idaho

A wonderland of cinder cones, lava flows, and volcanic features, all easily accessible to the casual walker. Volcanic features here are formed in lava very similar in composition to the lunar maria. The Visitor's Center is comprehensive and informative.

Geological Museum, Exhibition Road, London, England

A fine museum of geology with an excellent lunar display, showing large rocks and a nice panel display on the geological history of the Moon. Be sure to get a copy of the *Moon, Mars, and Meteorites* booklet while you are there; it has beautiful color graphics on planetary geology, and the minerals and dinosaurs are nice, too.

Hawaii Volcanoes National Park, Hawaii

The greatest of all the volcano parks, home to the active, nearly constant eruptions of Kilauea, a shield volcano that makes up the southern part of the big island of Hawaii. There are numerous fascinating lava features to see, including a complete, drained lava tube, similar to the sinuous rilles of the Moon. With luck, you will see an actual eruption!

Meteor Crater, Arizona, Visitor's Center

The world's first documented impact crater. If you happen to be traveling across country on Interstate 40 (Los Angeles to Chicago), stop off at Two Guns, Arizona (about 40 miles east of Flagstaff) to see this. It's a hole in the ground, but its historical significance is immense. The crater has a fine museum that describes the impact and how the crater was used to prepare the astronauts for their trips to the Moon. Get a copy of W. B. Hoyt's book *Coon Mountain Controversies*, and relive a century of scientific controversy.

National Museum of Natural History, Washington, D.C.

One of the Smithsonian's Mall museums, with a very good Moon display, several large rocks, and backlit panels describing lunar history. The unrivaled meteorite and mineral collection completes a "dry field" geological excursion.

National Air and Space Museum, Washington, D.C.

The holy-of-holies for the true space buff: Apollo spacecraft, mock-ups, memorabilia—and all the original (right) stuff! The best display is the "Apollo to the Moon" gallery; the lunar science display is so-so. (Be sure to go to the National Museum of Natural History across the Mall for its Moon display.) There is a fine gift shop, including a comprehensive book section.

Noordwijk Space Expo, ESTEC, Noordwijk, The Netherlands

The European Space Agency research center's visitor museum. The next time you find yourself in Holland, stop here. A variety of space exhibits, including a real Apollo Lunar Module. There is a lunar sample exhibit but not much on the Moon and its history. The Europeans are making noises that they want to go to the Moon early in the next century; we'll see.

Space Center Houston, Johnson Space Center, Houston, Texas

A new visitor's center, designed for hands-on activities. The center has an excellent display of the *Apollo 17* Command Module, a startlingly realistic lunar diorama including astronauts and rover, and an incredibly detailed mock-up (even down to the floor tiles!) of the Lunar Curatorial Laboratory, where the Moon rocks are kept. A tram tour takes you to one of two of the last flight models of the *Saturn 5* (now rusting in the Texas sun), to world-famous Mission Control, and to a variety of Space Shuttle and Space Station mock-ups. The gift shop is excellent for space toys and T-shirts but not as good for books.

Spaceport USA, Kennedy Space Center, near Cocoa, Florida

One of the best of the NASA visitor's centers. This one includes a real *Saturn 5*, the huge Vehicle Assembly Building, and a Lunar Module. Be sure to take the tour of the Cape's Air Force Test Range, which will take you by the decrepit and rusting Titan and Atlas launch pads, where space history was made 30 years ago. If you time it right, you can see a Shuttle launch. As at the Houston center, the gift shop has many space toys and T-shirts but a poor selection of books.

Sunset Crater National Monument, near Flagstaff, Arizona

A beautifully preserved cinder cone and lava flow, with a short trail that allows you to examine volcanic geology close-up. Dark ash that covers the surrounding area is quite similar to the dark mantle deposits found on the Moon. The field of explosion craters was used by the U.S. Geological Survey to train astronauts for their missions to the Moon; 30 years later, the area is used by local trail-bike riders.

Glossary
People, Places, and Terms

People

Entries are abbreviated and limited to people's contributions to our story. All Apollo astronauts are American; other nationalities are given where known.

Albritton, Claude. American geologist who published paper in 1935 (with Boon) advocating impact origin for very large, eroded structures on Earth.

Aldrin, Buzz. *Apollo 11* Lunar Module pilot; second man to walk on the Moon, July 1969; also flew on *Gemini 12*.

Archimedes. Ancient Greek mathematician and scientist who discovered principle of the lever and specific gravity (third century B.C.).

Aristarchus. Ancient Greek astronomer who calculated distance between Earth and the Moon (third century B.C.).

Armstrong, Neil. *Apollo 11* commander; first man to walk on the Moon, July 1969; also flew on *Gemini 8*.

Augustine, Norman. American chairman of presidential commission to examine future of American space program, 1990; CEO of Martin-Marietta Corp. (now Lockheed-Martin).

Baldwin, Ralph. American astronomer who worked out the impact origin of craters and the volcanic origin of maria in his book *The Face of the Moon* (1949).

Barringer, Daniel. American mining engineer who persistently advocated the impact origin of Meteor Crater, Arizona, in the early 20th century.

Bean, Alan. *Apollo 12* Lunar Module pilot; fourth man to walk on the Moon in November 1969; currently a space artist.

Boon, John. American geologist who published a paper in 1935 (with Albritton) advocating the impact origin for very large, eroded structures on Earth.

Cernan, Eugene. *Apollo 17* commander; 11th man to walk on the Moon and the last man to leave its surface, December 1972; also flew on *Gemini 11* and *Apollo 10*.

Conrad, Pete. *Apollo 12* commander; third man to walk on the Moon, November 1969; also flew on *Gemini 5* and *11* and *Skylab 2* (the first manned Skylab flight).

Copernicus, Nicolaus. Polish astronomer who worked out the sun-centered model of the solar system; the model was published after his death in 1543.

Darwin, George. English astronomer who developed the fission model for the origin of the Moon, in 1879.

Dietz, Robert. American geologist who developed the impact model for lunar and Earth craters (1946) and for Sudbury basin; also developed a model for the spreading of the sea floor, an essential element of plate tectonics on Earth.

Duke, Charles. *Apollo 16* Lunar Module pilot; 10th man to walk on the Moon, April 1972.

Duke, Michael. American geologist; former Lunar Sample Curator and joint originator (with Mendell) of the lunar base movement (the "lunar underground") in the 1980s.

Gagarin, Yuri. Russian cosmonaut; first man in space, April 1961; killed in an air accident in 1968.

Galilei, Galileo. Italian scientist who made the first observations of the Moon through a telescope, 1610.

Gilbert, Grove Karl. American geologist who published the first paper to analyze the Moon as a geological body in 1893.

Grimaldi, Francesco. Italian astronomer who published an early map of the Moon (1651) that Riccioli used for his system of naming features.

Gruithuisen, Franz von Paula. German astronomer (1830s) who first advocated the impact origin of craters on the Moon; also "discovered" a walled city on the Moon.

Herschel, John. British astronomer who was an unwitting participant in the "great Moon hoax" of the *New York Sun* newspaper in 1835.

Herschel, William. British astronomer in the 18th century; father of John Herschel; discoverer of the planet Uranus.

Hevelius. Polish astronomer (Johannes Höwelcke); early mapper of the Moon (1647).

Hooke, Robert. British scientist who made many discoveries and modeled craters on the Moon as burst bubbles in a boiling liquid (1689).

Irwin, James. *Apollo 15* Lunar Module pilot; eighth man to walk on the Moon, July 1971; later founded an evangelical ministry.

Kennedy, John F. American president who committed the nation (May 1961) to a manned lunar landing by 1970.

Langrenus. Flemish astronomer (Michel Florent van Langren) who drew the first map of the Moon with crater names (1645).

Laplace, Pierre Simon. French scientist who developed the nebular model for the origin of the solar system, in 1796.

McCauley, John. American geologist who coauthored (with Wilhelms) a major geological map of the Moon in 1971.

Mendell, Wendell. American astronomer and joint originator (with Michael Duke) of the lunar base movement (the "lunar underground") in the early 1980s.

Mitchell, Edgar. *Apollo 14* Lunar Module pilot; sixth man to walk on the Moon, January 1971.

Newton, Isaac. English physicist who formulated the law of gravitation and invented the calculus in his *Principia* (1687).

Nozette, Stewart. American planetary scientist who developed the original concept, in 1989, for the Clementine mission to the Moon, flown in 1994.

Plutarch. Greek author and philosopher (first century A.D.) who postulated that the Moon was another world, possibly inhabited.

Proctor, Richard. British astronomer who advocated the impact origin for lunar craters, in 1876.

Ptolemy. Greek astronomer (second century A.D.) whose Earth-centered model of the solar system was used for over a thousand years.

Regeon, Paul. American engineer; project manager for the Clementine mission to the Moon, 1994.

Rheita. Czech astronomer (Anton Maria Schyrleus of Rheita) who mapped the Moon in 1645; built Johannes Kepler's telescope.

Riccioli, Giambattista. Italian astronomer who developed the naming system for lunar features that is used today; used lunar map of Grimaldi (1651).

Rustan, Pedro. American engineer who was the manager of the Clementine program, 1994.

Schmitt, Harrison (Jack). *Apollo 17* Lunar Module pilot and geologist; 12th man to walk on the Moon, December 1972.

Scott, David. *Apollo 15* commander; seventh man to walk on the Moon, July 1971; also flew on *Gemini 8* and *Apollo 9*.

See, Thomas Jefferson Jackson. American astronomer (19th century) who advocated the capture model for the origin of the Moon.

Shepard, Alan. First American in space, *Freedom 7*, May 1961, and the *Apollo 14* commander; fifth man to walk on the Moon, January 1971.

Shoemaker, Carolyn. American astronomer; discoverer of comets and asteroids; codiscoverer of the Shoemaker-Levy comet, which hit Jupiter in July 1994.

Shoemaker, Eugene. American geologist; founder of planetary geology; invented lunar stratigraphic system; proved impact origin of Meteor Crater; codiscoverer of the Shoemaker-Levy comet, which hit Jupiter in July 1994.

Slayton, Donald K. ("Deke"). One of the original seven U.S. astronauts; grounded by a heart condition but chose the crews for all of the Apollo missions; finally flew in space as pilot of the Apollo-Soyuz mission in 1975.

Stafford, Thomas. American astronaut; chairman of the Synthesis Group (1990), a panel examining the Space Exploration Initiative; flew on *Gemini 6;* commanded *Gemini 9, Apollo 10,* and *Apollo-Soyuz.*

Teller, Edward. American physicist; inventor of the hydrogen bomb; early advocate of the Strategic Defense Initiative (1983) and "Brilliant Eyes," the sensor development program that became the Clementine mission.

Urey, Harold. American chemist; discoverer of heavy hydrogen; advocate of the scientific exploration of the Moon; wrote book *The Planets* (1952).

Verne, Jules. French author who wrote, in the 1870s, mostly scientifically accurate accounts of lunar flights.

Von Braun, Wernher. German physicist who developed the V-2, the world's first ICBM (1944); later developed Saturn rocket vehicles for the American space program (1960s).

Wegener, Alfred. German geologist who was an early advocate of continental drift; also wrote vigorous defense of the impact origin of lunar craters (1921).

Wells, Herbert George. British writer who wrote many science fiction stories, some dealing with space travel and the Moon (*First Men in the Moon,* 1901).

Wilhelms, Don. American geologist who made a global map of the Moon (with McCauley) in 1971 and definitively summarized lunar knowledge in his book *The Geologic History of the Moon* (1987).

Wood, Lowell. American physicist involved in the Strategic Defense Initiative; a disciple of Teller and cooriginator of the "Brilliant Eyes" concept, which later became the Clementine mission to the Moon (1994).

Wright, Orville. American inventor; brother of Wilbur; the first man to fly in a machine-powered aircraft, in December 1903.

Young, John. *Apollo 16* commander; ninth man to walk on the Moon, April 1972; also flew on *Gemini 3, Gemini 10, Apollo 10,* and two Space Shuttle flights, including the first (April 1981).

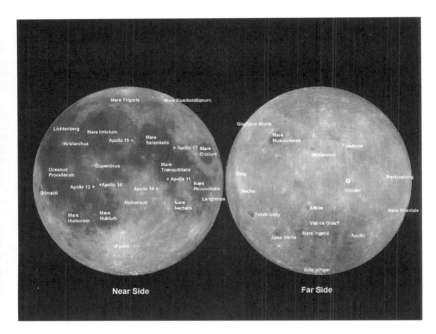

Figure GL.1. Map of the Moon showing the locations of some selected features and the locations of the *Apollo* landing sites. For detailed study, a full-sized map of the Moon should be obtained (see the bibliography).

Places

Please refer to the map above for locations of features on the Moon.

Lunar Craters

Albategnius (11.2° S, 4.1° E; 136 km).   Large crater in the central highlands, sketched by Galileo in 1610 (see Fig. 1.1).

Alphonsus (13.4° S, 2.8° W; 119 km).   Old crater with three dark, volcanic cinder cones on its floor.

Archimedes (29.7° N, 4.0° W; 83 km).   Crater flooded by mare basalt, demonstrating time span between Imbrium basin and its mare fill.

Aristarchus (23.7° N, 47.4° W; 40 km).   Very fresh crater excavating highland debris from beneath mare basalt cover (see Plate 8).

Compton (56.0° N, 105.0° E; 160 km). Small central-peak-plus-ring basin near lunar north pole.

Cone (3.5° S, 17.5° W; 370 m). Small, fresh crater excavating Fra Mauro breccias that were sampled on the *Apollo 14* mission, 1971.

Copernicus (9.7° N, 20.0° W; 93 km). Relatively young impact crater south of Mare Imbrium; defines Copernican stratigraphic system.

Descartes (11.7° S, 15.7° E; 48 km). Old crater in central highlands, near the landing site of *Apollo 16*, 1972.

Eratosthenes (14.5° N, 11.3° W; 58 km). Unrayed crater near Mare Imbrium; defines Eratosthenian stratigraphic system.

Flamsteed P (3.0° S, 44.0° W; 112 km). Old crater flooded by some of the youngest (1 billion years) lavas on the Moon; site of the *Surveyor 1* landing in 1966.

Fra Mauro (6.0° S, 17.0° W; 95 km). Old crater covered by ejecta from the Imbrium impact basin; near the landing site of *Apollo 14*, 1971.

Herigonius (13.3° S, 34.0° W; 15 km). Small crater north of Mare Humorum, near some of the most spectacular sinuous rilles in the maria.

Hortensius (6.5° N, 28.0° W; 15 km). Small crater, near which occur many small lunar shield volcanoes.

Kopff (17.4° S, 89.6° W; 42 km). Unusual crater, long thought to be volcanic, in Orientale basin; may have been created by an impact into a semi-molten melt sheet.

Lamont (5.0° N, 23.2° E; 175 km). Ridge ring system in Mare Tranquillitatis, formed over a two-ring basin.

Letronne (10.6° S, 42.4° W; 120 km). Crater largely flooded by mare basalt in Oceanus Procellarum.

Lichtenberg (31.8° N, 67.7° W; 20 km). Rayed crater that is partly covered by a very young mare lava flow, possibly less than 1 billion years old.

Linné (27.7° N, 11.8° E; 2 km). Very fresh, bright crater in Mare Serenitatis, reported before the space age to appear and disappear.

Ritter (2.0° N, 19.2° E; 29 km) and Sabine (1.4° N, 20.1° E; 30 km). Unusual twin impact craters in Mare Tranquillitatis, similar in morphology to Kopff.

Shorty (20.0° N, 31.0° E; 110 m). Small impact crater at the *Apollo 17* landing site, 1972; excavated dark mantle ash from beneath a layer of highland debris.

Sulpicius Gallus (19.6° N, 11.6° E; 12 km). Crater near a large exposure of dark mantle deposits.

Theophilus (11.4° S, 26.4° E; 100 km). Large crater on the edge of Mare Nectaris.

Tsiolkovsky (20.4° S, 129.1° E; 180 km).   Spectacular, mare-filled crater on the lunar far side.

Tycho (43.3° S, 11.2° W; 85 km).   Fresh, prominent rayed crater on the near side of the Moon; rays extend across entire hemisphere (Fig. 2.1); central peak exposes deep-seated rocks (see Plate 2).

Van de Graaff (27.0° S, 172.0° E; 234 km).   Double crater on the far side; site of a major geochemical anomaly caused by its location just inside the rim of South Pole–Aitken basin.

Lunar Maria

Mare Crisium (10–25° N, 50–70° E).   Mascon mare near the east limb; low to very low titanium basalts, extruded around 3.4 billion years ago.

Mare Fecunditatis (5° N–20° S, 40–60° E).   Complex, shallow mare made up of low-, moderate-, and high-titanium basalts, extruded about 3.4 billion years ago.

Mare Humorum (18°–30° S, 31–48° W).   Mascon mare on the southwestern near side of the Moon, filled with moderately high-titanium basalts, 3.2–3.5 billion years old.

Mare Imbrium (15–50° N, 40° W–5° E).   Mascon mare on the near side, deeply filled with low- and high-titanium basalts; age: from 3.3 to less than 2 billion years old.

Mare Nectaris (10–20° S, 30–40° E).   Mascon mare on the central near side; low-titanium basalts covering very high titanium basalts; age: 3.8–3.5 billion years.

Mare Nubium (10–30° S, 5–25° W).   Complex, shallow mare; low- and high-titanium lava flows; age: 3.3–3.0 billion years (?).

Mare Serenitatis (15–40° N, 5–20° E).   Mascon mare; very high titanium lavas around the margins and center of very low titanium lava; age: 3.8–3.3 billion years

Mare Smythii (5° N–5° S, 80–95° E).   Mascon mare, very shallow; moderate-titanium lava; possibly extremely young (1–1.5 billion years).

Mare Tranquillitatis (0–20° N, 15–45° E).   Complex, shallow, irregular mare; site of the first lunar landing; old (3.8 billion years), with very high titanium lavas

Oceanus Procellarum (10° S–60° N, 10–80° W).   Complex, shallow, irregular mare; largest on the Moon; many compositions, with ages including the youngest lavas on the Moon (less than 1 billion years old).

Sea of Rains. *See* Mare Imbrium.

Sea of Tranquillity. *See* Mare Tranquillitatis.

Sinus Medii (3° S–5° N, 5° W–5° E). Small patch of mare near the exact center of the lunar near side; site of the *Surveyor 6* landing (1967).

Lunar Basins

Crisium basin (17.5° N, 58.5° E; 740 km). Nectarian-age multiring basin; ejecta possibly sampled by the *Luna 20* mission.

Humorum basin (24° S, 39.5° W; 820 km). Nectarian-age basin south of Procellarum.

Imbrium basin (33° N, 17° W; 1,150 km). Major large basin on the Moon; defines base of Imbrian System; formed 3.84 billion years ago; its ejecta was the sampling objective of the *Apollo 14* and *15* missions.

Nectaris basin (16° S, 34° E; 860 km). Defines base of Nectarian System; possibly sampled on the *Apollo 16* mission in 1972; age: 3.92 billion years.

Orientale basin (20° S, 95° W; 930 km). Youngest large, multiring basin on the Moon, formed sometime after 3.84 billion years ago; its interior and exterior deposits were used as a guide to interpret older, degraded basins.

Procellarum basin (26° N, 15° W; 3,200 km). Alleged impact basin, supposedly the largest on the Moon; *Clementine* laser altimetry data do not support its existence.

Schrödinger (75.6° S, 133.7° E; 320 km). Type example of a two-ring basin, near the south pole of the Moon; formed after Imbrium basin but before Orientale basin.

Serenitatis basin (27° N, 19° E; 900 km). Nectarian-age multiring basin, sampled and explored by *Apollo 17* mission in 1972; age: 3.87 billion years.

South Pole–Aitken basin (56° S, 180°; 2,500 km). Largest, deepest (over 12 km) impact crater known in the solar system; oldest basin on the Moon; absolute age unknown (4.3 billion years??).

Other Lunar Surface Features

Apennine Bench (25–28° N, 0–10° W). Refers to a relatively elevated region near Archimedes and just inside the rim of Imbrium basin; includes light-toned, nonmare KREEP volcanic lava plains, called the Apennine Bench Formation.

Apennine Mountains (15–30° N, 10° W–5° E). Large mountain chain making up the southeastern rim of the Imbrium basin.

Cayley plains. Light-toned, smooth highland plains, first defined in the central near side but having moonwide distribution; probably a form of impact ejecta from the youngest major basins; may cover ancient mare lavas in some areas.

Cordillera Mountains (10–35° S, 80–90° W). Arcuate mountain chain that makes up the rim of the Orientale basin.

Hadley-Apennines (26° N, 4° W). Informal name given to the region of the *Apollo 15* mission exploration; includes mare, Hadley Rille, and Apennine highlands.

Hadley Rille. *See* Rima Hadley.

Marginis swirls (15° N, 90° E). Light-toned swirls north of Mare Marginis; origin unknown.

Marius Hills (10–15° N, 50–60° W). Complex area of small domes, cones, and sinuous rilles in Oceanus Procellarum; the dome-like swell may indicate that this region is a large lunar shield volcano.

Reiner Gamma (7° N, 59° W). Bright, swirl-like deposit in Oceanus Procellarum; origin unknown.

Rima Bode II (13° N, 4° W). Cleftlike vent and linear trench outline vent system for a large, regional blanket of dark volcanic ash.

Rima Hadley (25° N, 3° W). Long sinuous rille starting in the highlands and emptying into the maria; probably a lava channel and/or tube.

Rümker Hills (41° N, 58° W). Complex of cones and domes in Oceanus Procellarum, similar to Marius Hills but much smaller.

Taurus-Littrow (20° N, 31° E). Informal name given to the region of the *Apollo 17* mission exploration; includes mare, dark mantle, and the highlands of the Serenitatis basin.

Tranquillity Base (1° N, 23° E). Site, in Mare Tranquillitatis, of man's first landing on the Moon, *Apollo 11*, July 20, 1969.

Earth Impact Structures

Chicxulub structure (21.3° N, 89.6° W; 250–300 km). Large, multiring basin in the northern Yucatan, formed 65 million years ago; the crater responsible for the Cretaceous-Tertiary mass extinction (demise of the dinosaurs).

Meteor Crater (35° N, 111° W; 1200 m). Simple, bowl-shaped crater in Arizona; first documented impact structure on Earth, formed about 20,000 years ago.

Sudbury structure (46 5° N, 81° W; 200–250 km). Large rock province in central Ontario formed about 1.5 billion years ago; may have made a very large sheet of impact melt that segregated into many different rock types.

Terms

The definitions given here are not exhaustive and are oriented to the use of these terms in this book.

absolute age.   The age in years of a geological rock unit.

absolute zero.   The temperature at which the motion of all atoms and molecules stops (-273 °C).

absorption band.   In a spectrum, a zone of relatively low reflectance, caused by the presence of a mineral.

accretion.   The assembly of a larger body from many smaller bodies; all planets and moons form by accretion.

agglutinate.   An impact melt made up of glass, mineral, and rock fragments.

aggregate.   A rock made up of many different fragments of other rocks.

albedo.   The reflectivity of an object; light and dark are high and low albedo, respectively.

alkali.   A series of light metal elements, including sodium and potassium.

ALSEP.   Acronym for Apollo Lunar Surface Experiment Package, the network station set up on the Moon by the last five Apollo landings.

aluminous melt group.   Impact-melt breccias that contain large amounts of aluminum.

aluminum.   A metal element abundant on the Moon; atomic number 13, atomic weight 26.9, symbol Al.

amorphous.   Having no internal structural order; glassy.

angular momentum.   The product of angular velocity and mass; a property of rotating systems.

annealing.   Recrystallization of minerals in the solid state.

anomalies.   Something markedly unexpected or a value different from many others in a data sequence.

anorthosite.   A slowly cooled rock made up almost solely of the calcium-rich mineral plagioclase feldspar.

Antares.   The *Apollo 14* Lunar Module (named after the star in the constellation Scorpio).

antipode.   The point on a globe directly opposite (180° on a great circle) a given location.

aperture.   The diameter of a telescope objective (lens or mirror).

apogee.   The highest altitude in an elliptical orbit around Earth.

Apollo.   Space program created in 1961 to land a man on the Moon and return him to Earth, a mission achieved in July 1969.

architecture. The plans, elements, strategies, and equipment needed to conduct some defined mission in space.

argon. A noble gas element; atomic number 18, atomic weight 39.9, symbol Ar; produced by the Sun and by the natural decay of radioactive potassium.

armalcolite. An iron, titanium oxide mineral, found in lunar mare basalts and named for the crew members of the *Apollo 11* mission.

array. A number of instruments (e.g., telescopes) used together in coordination as a single instrument.

ash. Very small fragments of lava sprayed out at a vent, cooled quickly, and deposited as a blanket of debris.

ash flow. Ash that behaves as a single, massive fluid; may travel great distances.

asteroid. A small body, usually less than a few hundred kilometers in size, that orbits the Sun as an independent planet.

astronaut. The human pilot or occupant of a spacecraft.

astronomical. Dealing with objects and phenomena in the sky, their observation, and the instruments used to observe them.

atomic number. The number of protons in an atomic nucleus.

atomic weight. The total number of protons and neutrons in a nucleus.

atrophy. Loss of muscle mass caused by long-term exposure to zero-g.

Augustine Committee. Presidential commission chartered in 1990 to examine the nation's space program.

axis. Imaginary line about which a moon or planet rotates; intersects the surface at the poles

basalt. A dark, fine-grained rock, rich in iron and magnesium, created by solidification of lava.

basin. A very large impact crater, usually greater than 300 km in diameter.

bedrock. The intact layer of rock below the regolith; makes up regional geological units of the planets.

binary accretion model. Holds that the Moon and Earth accumulated as separate, independent planets.

bombardment. The collision of a planet with asteroids, repeatedly over a long time.

booster. The first rocket stage that carries a spacecraft from Earth's surface toward orbit.

boulder. A rock, usually 1 m or larger in size.

breccia. A rock composed of angular pieces of other rock and mineral fragments.

bulk composition. The chemical or mineral composition of an entire planet or satellite; it cannot be measured directly but must be calculated.

calcium.   A light metal element, abundant in the Moon's crust; atomic number 20, atomic weight 40.1, symbol Ca.

caldera.   Summit crater of a volcano; collapse depression found over a subsurface magma chamber.

capture model.   Holds that the Moon formed elsewhere in the solar system and was captured by Earth's gravity into orbit.

carbon.   A light element, implanted into the regolith from the solar wind; atomic number 6, atomic weight 12, symbol C.

carbon dioxide.   Gas made up of one carbon, two oxygen atoms; formula CO_2.

carbon monoxide.   Gas made up of one carbon, one oxygen atom; formula CO.

cataclysm.   The concept that most or all of the large basins and craters on the Moon formed at a single time, 3.8–3.9 billion years ago.

catastrophe.   A geological event short in duration but widespread in effect.

central peak.   A mountain in the center of complex impact craters, derived from relatively great depths.

Challenger.   (1) *Apollo 17* Lunar Module, named for the British ship that charted the Antarctic in 1870; (2) Space Shuttle that exploded shortly after launch on January 28, 1986, killing all seven crew members.

chasm.   A large canyon or gorge on a planet's surface.

chemical analysis.   Measuring the amounts of the elements present in a planet's surface or in samples thereof.

chemical bond.   Connection between elements to make up a compound.

chemical composition.   The elemental makeup of a substance.

cinder cone.   A hill produced by the buildup of ash or other pyroclastic fragments around a volcanic vent.

circular maria.   An area of dark lava on the Moon; fills large circular basins.

clast.   A fragment of rock or mineral in a breccia.

Clementine.   Department of Defense mission in 1994 designed to map the Moon and a near-Earth asteroid; the Moon was completely mapped, but the asteroid portion of the mission was canceled.

coagulate.   The accumulation of solids in a liquid.

cobalt.   A heavy metal element; atomic number 27, atomic weight 58.9, symbol Co.

collapse pit.   A small depression associated with subsurface collapse, as over a shallow chamber after lava has been drained from it.

Collisional Ejection Hypothesis ("Big Whack").   Holds that the Moon formed when a Mars-sized planet collided with the early Earth, spraying material into orbit; this material later accumulated into the Moon.

cometary nucleus.   A small solid object, made up of dust and water ice, that occurs in the head of a comet.

Command-Service Module (CSM).   Spacecraft that remained in lunar orbit during the Apollo missions and returned the astronauts to Earth.

complex crater.   A crater with terraced walls, flat floors, and central structures; on the Moon these craters have diameters between 20 and 300 km.

compression.   Forcing mass into a smaller volume (increase of density).

concentrating mirror.   A concave mirror that can focus light into a small area, increasing its intensity.

concentric.   Circular elements that occur at different radii from one central area.

cone.   A small volcanic construct, usually having relatively steep sides.

conglomerate.   A rock made up of other rocks; distinguished from a breccia by its rounded clasts.

contiguous.   Having lateral or vertical continuity, as in regional rock units.

Copernican System.   Geological classification on the Moon encompassing the freshest, rayed craters and some other minor features that formed in the last 1 billion years.

core.   The central zone of a planet, usually made up largely of iron metal.

cosmic ray.   A very high energy nuclei of atoms, traveling at high speeds through space; hazardous to life.

cosmology.   The study of the origin of the universe.

crater.   A hole resulting from the collision of an object with a planetary surface.

Cretaceous-Tertiary boundary.   A time horizon 65 million years ago in Earth's geological record, corresponding to a mass extinction of life, including dinosaurs.

crust.   The outer zone of a planet, made up of relatively low density rocks.

crustal magmatism.   The formation of rocks by intrusion of magma into the crust.

crustal rock.   Rocks that make up the crust of a planet.

cryogenic.   Dealing with very low temperatures, where common gases are found in the liquid state.

crystal. A substance possessing permanent and regular internal order or structure.

crystallization age. The length of time since a rock or mineral formed from a liquid state (by cooling of magma).

crystallize. precipitation of minerals out of a cooling magma.

dark mantle deposit. A large area of the Moon covered by glassy ash deposits.

dark side. The nighttime hemisphere of the Moon; *not* the same as the far side.

Delta Clipper (DCX). An experimental rocket made by the Department of Defense in 1993 as a test bed for a single-stage to orbit launch vehicle.

dome. A bulbous, hill-like volcano.

drill core. Rocks extracted from a planet by drilling into the surface with a hollow tube to examine the vertical dimension of rock units.

dust clouds. Dust kicked up by the exhaust of a rocket engine, or the layer of dust levitated by an electrostatic charge near the lunar terminator.

Eagle. The *Apollo 11* Lunar Module.

ecliptic plane. The plane in which Earth orbits the Sun.

effusion. The eruption of flowing lava from a volcanic vent.

ejecta. Debris thrown out of an impact crater; by tradition, this Latin plural is considered singular in written English.

ejecta blanket. Debris that results from an impact and that surrounds the crater rim.

electron microscope. An instrument that uses X-rays to produce highly detailed, high-magnification images of objects and their surfaces.

element. A substance defined by number of protons and electrons, having definite and distinct chemical and physical properties.

elliptical orbit. A noncircular orbit shaped like an ellipse, having a low and high altitude.

equatorial orbit. Orbit of spacecraft above the equator of a planet.

equatorial plane. The imaginary plane defined by the equator of a planet or the Sun.

equilibrium. State in which all conditions remain constant with respect to each other, although the set as a whole may change.

Eratosthenian System. Geological system that includes relatively fresh craters and some mare lava flows with ages from 3.2 to about 1 billion years.

erosion. The destruction of planetary surface features by some process.

error ellipse. A zone on a planet at which a landing spacecraft may be expected to set down, allowing for potential errors in navigation and flight.

europium. A rare earth element, diagnostic of the early history of the Moon; atomic number 63, atomic weight 151.9, symbol Eu.

Explorer 1. The first successful U.S. Earth-orbiting satellite, launched in January 1958.

extension. The tensile stress of a planetary surface; results in tensional features (cracks, grabens).

extravehicular activity (EVA). The time an astronaut spends outside of a spacecraft.

extrusion. Lava that pours out of a planet's interior onto the surface.

far side. The hemisphere of the Moon opposite Earth, permanently out of view of observers on Earth; *not* the same as the dark side.

fault scarp. A cliff formed by the surface expression of a fracture in a planet's crust.

feldspar. A rock-forming mineral rich in calcium and aluminum.

ferroan anorthosite. A rock made up almost entirely of plagioclase feldspar, rich in aluminum and calcium; the principal rock type of the lunar crust.

fieldwork. The study of rocks and geological features in their natural environment.

fire fountain. A spray of lava from a vent, producing a dark mantle deposit.

fission model. Holds that the Moon split off from a very rapidly rotating, liquid early Earth.

flow front. The terminal end of a lava flow.

fluorine. A halogen gas element; atomic number 9, atomic weight 18.9, symbol F; used to produce oxygen by reduction process from highland materials.

flyby. A mission technique whereby a planet is examined by a spacecraft as it flies by the planet on the way to another destination.

fossil fuel. Petroleum-based energy substances, including oil and natural gas.

fracture. A zone of failure in a geological material.

fragmental breccia. A breccia containing little or no melt, made up of fragments of many other rocks.

Freedom 7. First U.S. manned spacecraft, piloted on a suborbital flight by Alan Shepard on May 5, 1961.

fusion. The joining together of the nuclei of two different atoms, resulting in energy release; the cause of the Sun's energy output.

galaxy. A collection of stars held together by gravity.

gamma radiation. High-energy radiation produced during nuclear reactions.

gamma-ray spectrometer. An instrument that measures gamma radiation as a function of energy; can measure chemical composition of materials and planetary surfaces.

gaseous sodium. A low-density cloud of sodium atoms; the principal component of the lunar atmosphere.

gaseous sulfur. A low-density cloud of sulfur atoms.

Gemini. American space program of two-man spacecraft flight tests in 1965–66; the precursor to the Apollo program.

Geographos. Near-Earth asteroid. the target of the Clementine mission; the flyby was canceled after spacecraft failure on May 3, 1994.

geological history. The natural evolution and history of a planet, reconstructed though study of its rocks and surface features.

geological past. Typically, the distant past, meaning millions or billions of years ago.

geological unit. A body of rock or fine material that has a common origin, has lateral and vertical continuity, and deposited at or over a certain time.

geology. The study of the processes and history of the solid, rocky objects of the solar system.

geophone. A small seismometer, designed to be used in an array with others, to decipher the subsurface structure of a planet.

geophysics. The study of the subsurface and whole-body nature of the planets, including their interaction with the space environment.

geostationary orbit. Equatorial orbit 22,000 miles above Earth, during which the orbital period equals one day (24 hours); as a result, the spacecraft appears to "hover" over one point on Earth.

glass. A natural material that possess no internal order; a liquid of high viscosity.

global field. A magnetic field that surrounds a whole planet; generated by motions within a core of liquid metal.

global figure. The detailed shape of a planet.

graben. Two parallel, inwardly dipping faults with a down-dropped block between them.

granulite. A rock that has recrystallized in the solid state under intense heat.

granulitic breccia. A breccia that has been partly recrystallized in the solid state under intense heat.

gravity. The mutual attraction of objects that have mass.

gravity map. A map showing the variation in gravitational attraction across the surface of a planet, caused by variations in the internal density of a planet.

halogen. Chemically related, nonmetallic elements that include fluorine, chlorine, and bromine.

hard lander. A spacecraft designed to collect information and then impact into a planet; may be designed to partly survive the landing.

helium. A light noble gas emitted naturally from the Sun; atomic number 2, atomic weight 4, symbol He.

helium-3 (^{3}He). A helium isotope that contains only one neutron and that can be fused with deuterium to produce energy; rare on Earth but found on the Moon among the solar gases of the regolith.

high-energy particles. Small nuclei of atoms that travel at very high speeds; produced during stellar explosions.

highlands. The light-toned, heavily cratered uplands of the Moon; synonym of terrae.

host material. A substance that encloses another rock or mineral; the matrix of a breccia.

hydrated mineral. A mineral containing water molecules within its crystal structure.

hydrogen. The lightest and most abundant element in the universe; atomic number 1, atomic weight 1, symbol H; main component of solar wind.

hydrothermal. Of or pertaining to the geological effects of very hot, mineral-laden waters; often associated with ore deposits on Earth.

hydrous phase. A substance containing water.

hypothesis. An assumption or imagining that is tested by experiment or observation.

ICBM. Acronym for Intercontinental Ballistic Missile, the general term for a rocket booster capable of delivering a nuclear warhead or launching spacecraft.

igneous. Of or relating to material formed from a liquid state.

igneous rock. A rock crystallized from a liquid (magma).

ilmenite. An iron and titanium mineral found in mare basalts.

Imbrian System. Geological system made up of large craters, basins, and many lava flows; includes units emplaced between 3.84 and 3.2 billion years ago.

impact. Geological process resulting from the collision of objects in space.

impact dark-halo crater. A crater with dark ejecta, caused by the exhumation of buried lava flows in the lunar highlands.

impact flux. The number of impacts as a function of time.

impact melt. The portion of the impact target that is shocked to such high levels that the rocks completely melt; lines the cavity of the growing crater.

impact-melt breccia. A rock made up of broken mineral and rock fragments, cemented together by impact melt.

impactor. The object that strikes a planetary target; an asteroid, comet, or meteorite.

infrared. Light having wavelengths greater than red light (> 1 λm).

interferometer. A collection of instruments designed to operate as a single instrument; uses the principle of interfering light or radio waves to resolve distant objects.

International Space Station. The current name for NASA's project to build a permanently manned space station in Earth orbit.

ionic radius. The size of an atom; the determinant, along with the charge of an atom, of whether or not an atom can enter the crystal structure of a mineral.

ionosphere. The zone of charged particles above Earth's atmosphere, trapped by lines of the global magnetic field.

iridium. A metal element, found in meteorites; atomic number 77, atomic weight 192.2, symbol Ir.

iron. A very common metal and rock-forming element; atomic number 26, atomic weight 55.8, symbol Fe.

isotope. A variant of an element; caused by an excess or deficiency of neutrons.

isotopic composition. The amounts of each isotope of an element in a rock or geological sample.

KREEP. Acronym for potassium (K), rare earth elements (REE), and phosphorous (P); a chemical component in lunar rocks; created as the last phase of the magma ocean.

krypton. A noble gas element; atomic number 36, atomic weight 83.8, symbol Kr.

laser altimetry. The determination of the topography of a planet by measuring the time it takes laser pulses to travel round trip between the spacecraft and the planet's surface.

laser reflector. Mirrors placed on the Moon by the Apollo astronauts to allow the Earth-Moon distance to be measured very precisely.

lava. Liquid magma extruded onto a planetary surface.

lava channel. A channel filled with a stream of flowing lava or the landform produced after such a process.

lava tube. A lava channel that is partly or completely roofed over to enclose the lava stream; may form a cave after the flow has cooled.

law of superposition. Older geological units overlaid or intruded by younger geological units; the fundamental principle of stratigraphy.

lead. A heavy metal element; atomic number 82, atomic weight 207.2, symbol Pb.

leading edge. The edge on the rising side of an object in the sky.

libration. A slight movement in latitude and longitude that allows us to see more than 50 percent of the lunar surface.

limb. The edges of the Moon's apparent disk in the sky (the eastern and western 90° longitudes).

lineaments. The linear alignment of small features on planetary surfaces.

LKFM. Acronym for "low-K Fra Mauro," a variety of lunar impact melt that has a composition richer in iron and magnesium than the upper crust.

lobate scarp. Small cliffs that have a lobe shape; found at the termini of lava flows.

low Earth orbit (LEO). Orbit around Earth between 100 and 200 km; typically used as a parking orbit before interplanetary departure.

Luna. (1) The Latin name for Earth's Moon; (2) Soviet program of robotic spacecraft made up of flybys, orbiters, and landers, including three sample-return missions.

Lunakhod. Soviet robotic lunar surface rovers (1970 and 1973), remotely controlled from Earth.

lunar day. The time it takes for the Moon to rotate once on its axis; equal to 29.5 Earth days or about 709 hours.

lunar eclipse. Occurs when Earth gets between the Moon and the Sun; always happens at full moon phase.

Lunar Module. The Apollo spacecraft (1969–72) designed to land two men on the Moon.

Lunar Orbiter (LO). American robotic mission series (1966–67) that orbited five spacecraft and photographed Apollo landing sites and general scientific targets.

Lunar Polar Orbiter (LPO). A proposed mission (1973) to map the Moon globally with a variety of remote-sensing techniques; never flown.

Lunar Roving Vehicle (LRV). Surface vehicle used by the crews to traverse the Moon on the last three Apollo landing missions (1971–72); also referred to as "the rover."

lunar thermal history. The amounts of heat produced and lost by the Moon as a function of time; reconstructed from its rock record.

lunar transient phenomena (LTP). Unusual lights or glowing clouds on the Moon, occasionally observed from Earth.

mafic.   Geological property describing an enrichment in iron and magnesium at the expense of silicon and aluminum.

magma.   Liquid rock within the interior of a planet.

magma electrolysis.   The production of oxygen by passing an electrical current through a body of melted rock or soil.

magma ocean.   The state of the early Moon in which the entire globe was covered by a layer of liquid rock hundreds of kilometers thick.

magnesium.   A light metal element; atomic number 12, atomic weight 24.3, symbol Mg.

magnetic anomaly.   A zone of a planetary surface where the magnetic field is either stronger or weaker than expected; any zone of intense magnetization on the Moon.

magnetic field.   A region in which a magnetic force is detectable everywhere.

magnetized.   Material that is surrounded by lines of magnetic force; it may attain this magnetism by passing through an existing field or cooling in the presence of such a field.

magnetosphere.   A zone surrounding Earth and containing charged particles trapped by lines of Earth's magnetic field.

mantle.   The part of a planet below the crust and above the core; after partial melting, the source of magma in most planets.

maria (MAR-ee-ah).   The dark areas of the Moon (the Latin plural for "seas"; the singular is *mare* [MAR-ay]).

mascons.   Acronym for "*mass con*centrations," a zone of anomalously high density within the Moon; detected over the circular mare basins.

massif.   A large, discrete mountain.

matrix.   The breccia background ("groundmass") in which the clasts are set.

maturation.   The development of agglutinates in the regolith; darkens the soil and suppresses spectral features caused by minerals.

megaregolith.   The broken-up, impact-processed, outer few kilometers of the lunar crust.

melt sheet.   The zone of shock-melted rock that lines the floor of an impact crater.

Mercury.   American space program of the first manned spaceflights, 1961–63.

metamorphic rock.   A rock recrystallized in the solid state under high temperatures and pressures.

meteor.   A rock or ice particle entering Earth's atmosphere at cosmic velocities (greater than several km/sec), forming a tail of ionized gas.

meteorite.   A rock found or observed to fall from the sky, derived from extraterrestrial planetary bodies.

meteoroid. Any small, natural object (rock or ice) traveling through space.

Mg-suite. A series of related igneous rocks that make up a significant fraction of the crust of the Moon, distinct from the ferroan anorthosites.

microgravity. "Zero-g," or the free fall experienced by orbiting objects.

micrometeorite. A dust particle traveling through space at high speeds.

micron. One-millionth of a meter, or one-thousandth of a millimeter; officially, "micrometer."

microwave. Very high frequency radio waves (cm-scale wavelength).

mineral. A naturally occurring substance having a crystalline structure; rocks are aggregates of minerals.

mineralogy. The study of minerals.

model. A set of descriptors or rules that may or may not have a counterpart in nature; constructed by scientists to try to understand complex phenomena.

molten. Being in the liquid state.

molybdenum. A heavy metal element; atomic number 42, atomic weight 95.9, symbol Mo.

moon. A generic term for a natural satellite of a planet.

Moon. The natural satellite of Earth, also called *Luna*.

moonquake. A natural seismic event on the Moon.

morphology. The shape of something; as applied to geology, the study of landforms and how they are created and evolve (geomorphology).

multiring basin. A large feature of impact origin having a multiple-, concentric-ring structure.

multispectral map. A map of a planet or region that shows the distribution of units of color, inferred to relate closely to surface composition.

N-1. Soviet "superbooster" of the late 1960s, designed to carry the lunar spacecraft; destroyed by catastrophic explosions on two occasions, causing the Soviets to fail to land a man on the Moon.

NASA. The National Aeronautics and Space Administration, America's "space agency" (founded 1958).

natural radioactivity. The spontaneous decay of one element into another (for example, uranium naturally turning into lead over several billion years).

near side. The hemisphere of the Moon that constantly faces Earth.

nebulae. Large clouds in space in which new stars and planets are created.

nebular material. The gas and dust from which solar systems are made.

nebular model. Holds that our solar system was created out of a rotating cloud of gas and dust; developed by Laplace (1796).

Nectarian System. Geological system that includes crater and basin deposits and some ancient maria; includes units deposited between 3.92 and 3.84 billion years ago.

neon. A noble gas element; atomic number 10, atomic weight 20.2, symbol Ne.

nickel. A heavy metal element; atomic number 28, atomic weight 58.7, symbol Ni.

nitrogen. A light gas element; atomic number 7, atomic weight 14, symbol N.

norite. An igneous rock of the lunar crust; made up of approximately equal amounts of plagioclase feldspar and magnesium-rich pyroxene minerals.

nuclear fusion. The joining together of the nuclei of two elements to make a third element, releasing energy in the process; what makes the stars shine.

olivine. A magnesium- and iron-rich silicate mineral, common in mafic igneous rocks.

orbit. The condition of balance between forward velocity and the free fall of an object toward a planet; stable, unless acted on by another force.

Orbital Transfer Vehicle (OTV). Hypothetical spacecraft designed to transport payloads from the LEO space station to higher orbits, such as geostationary orbit.

orbiter. A spacecraft designed to orbit a planetary body; often refers to a spacecraft that collects planetary data by remote-sensing techniques.

ore. Any material that can be mined at a profit.

outcrop. A natural exposure of bedrock.

oxygen isotope. Oxygen atoms containing a number of neutrons different from the number of the most abundant isotope, ^{16}O; indicates a close relation between Earth and the Moon.

parallax. The slightly different lunar views available to an observer moving from side to side; permits us to partly "see around" the lunar far side.

payload. The cargo or object of value carried into space by a booster rocket.

perigee. The low point of an elliptical orbit around Earth.

phosphorous. A light, nonmetallic element; atomic number 15, atomic weight 30.9, symbol P.

plagioclase. A silicate mineral rich in aluminum and calcium; a subset of the feldspar group; common in anorthosite, the rock of the lunar highlands.

plains.   A relatively flat, smooth material that fills depressions on the Moon.

planetoid.   A small planetary object, such as a large asteroid.

plasma.   A low-density, high-energy form of matter that is the fourth state of matter (after solid, liquid, and gas).

pluton.   A body of rock solidified at depth within a planetary crust.

plutonic rock.   A rock from a pluton; often has a very large crystal size, caused by very slow rates of cooling.

polar orbit.   An orbit in which the planet's spin axis is within the orbital plane of the spacecraft (the orbital plane is fixed in space, and the entire planet will slowly rotate into the view of a spacecraft in such an orbit).

potassium.   A light metal element; atomic number 19, atomic weight 39.1, symbol K.

power law.   A mathematical relation in which two variables relate to each other exponentially; as applied to regolith particle and crater size distributions, the smallest members of the population greatly dominate the larger sizes.

pre-Nectarian.   In lunar geological time, the rocks laid down before 3.92 billion years ago, including many craters and basins and some maria.

pristine rock.   A lunar rock crystallized from an internally generated magma.

proto-Earth.   The early Earth as it was before the hypothetical large impact that created the Moon.

province.   A region of similar terrains or a collection of geological units with similar or related origins.

pyroclastic.   Literally "fire-broken," meaning fragmental rocks produced in explosive volcanic eruptions; includes ash.

pyroxene.   A magnesium- and iron-rich silicate mineral; an important constituent of mafic igneous rocks such as basalt.

quartz.   A mineral of the compound silicon dioxide (SiO_2), very common on Earth but extremely rare on the Moon.

radar.   Acronym for "*ra*dio *d*etection *a*nd *r*anging"; can be used to image planetary surfaces, producing information on physical properties and compositions.

radiation.   The radiant transfer of energy through space.

radiogenic isotope.   An element that naturally and spontaneously decays into another element.

radiogenic lead.   Lead created from the radioactive decay of uranium.

radiometric dating. Determining the age of the formation of rocks and minerals by measuring their amounts of radiogenic isotopes, which decay at known rates.

Ranger. American space program of robotic hard-landing spacecraft that took the first close-up pictures of the lunar surface in 1964 and 1965.

rays. Light-toned material radiating away from the ejecta blanket of the very freshest craters.

reconnaissance. The initial or cursory examination of a planet or a set of data.

Redstone. The booster rocket for the first manned Mercury space-flights (1961).

reflectance. The amount of light reflected off a body or object in space.

reflectance curve. The amount of light reflected from a planetary surface as a function of wavelength (color); characteristic of mineral composition.

regolith. The unconsolidated mass of debris that overlies bedrock on the Moon; created by impact bombardment.

relative age. The age of a geological unit with respect to other geological units (e.g., younger or older) without regard to its age in years.

remote sensing. The determination of the compositional or physical properties of planetary surfaces from a distance.

residuum. "What's left over," specifically, the last phases of liquid in the crystallization of magmatic systems (e.g., KREEP in the Moon).

resolution. The capability of a system to make clear and distinct the separate parts or components of an object.

retrograde. Motion or rotation in a direction opposite to the natural counterclockwise motion (viewed from the north pole) of solar system objects.

rille. A linear or sinuous (snakelike) depression on the Moon.

rim crest. The topographic high ground that surrounds a crater and that marks the rim.

rings. Concentric terrain elements that surround or lie within basins on the terrestrial planets.

robot. A machine capable of following programmed instructions to perform work.

rock. An aggregate of minerals; the main constituent of the terrestrial planets.

rockberg. A floating raft of freshly crystallized plagioclase in the magma ocean.

rotation. The spinning of an object about a central axis.

rover.   A manned or robotic machine designed to travel across a planetary surface.

rubidium-strontium dating.   A dating technique that uses isotopes of the elements rubidium and strontium (half-life 4 billion years) to date rocks.

samarium.   A rare earth element, component of KREEP; atomic number 62, atomic weight 150.4, symbol Sm.

Saturn 5.   American heavy-lift booster that took Apollo spacecraft to the Moon (1968–72); currently, lawn ornaments at NASA field centers.

secondary crater.   A small crater produced by the impact of a clot of debris thrown out from a larger crater.

sedimentary rock.   Rock formed from deposits of ground-up material (sediment).

seismic lines.   An array of geophones laid out to determine the subsurface structure of an area or region.

seismic profiling.   The detonation of explosive devices that create seismic events for deployed seismic lines to determine subsurface structure.

shatter cone.   A striated feature formed in fine-grained rocks by the passage of a shock wave; indicative of impact.

shield volcano.   A broad, low-relief volcanic construct made up of flows of relatively fluid lava, usually basalt.

shock melt.   Material melted by the passage of a shock wave.

shock pressure.   The amount of compression produced by the passage of a shock wave.

shock wave.   A wave characterized by very high pressures that last for extremely short periods of time ("megabars for microseconds").

sidereal period (month).   Time taken for one revolution of the Moon around Earth; 27.3 days or about 656 hours.

siderophile element.   An element that tends to follow iron and other metals (such as nickel) during igneous processes.

silica.   The short name for the compound silicon dioxide (SiO_2).

silicate.   A mineral group in which the silicon tetrahedron (a single silicon atom surrounded by four oxygen atoms, SiO_4) is always part of the structure; the largest mineral group, also called the "rock-forming" minerals.

silicon.   A very common light element; atomic number 14, atomic weight 28.1, symbol Si.

simple crater.   An impact feature characterized by a bowl-shaped hole.

sintering.   The welding together of rock and mineral fragments to create solid rock; a process that could be used on the Moon to make bricks, by sintering soil in a solar thermal furnace.

sinuous rille. A lava channel or tube in the lunar maria.

Skylab. American space program (1973–74); flew a manned laboratory in Earth orbit in 1973, which fell to Earth in 1979.

sodium. A light metal element; atomic number 11, atomic weight 22.9, symbol Na.

soil. Fine material that overlies bedrock; similar to regolith but without the coarse, large rock fragments.

solar cell. A device that uses the silicon photoelectric effect to produce electrical energy in space from sunlight.

solar eclipse. An eclipse caused when the Moon moves between the Sun and Earth; always occurs at new moon.

solar panel. An array of solar cells, used to provide electrical energy in space.

solar thermal furnace. A concentration of solar energy with a mirror or lens; used to melt or sinter material, for example, soil into brick.

solar wind. A stream of gases, mostly hydrogen, emanating from the Sun; responsible for implanting gas into the lunar regolith.

space age. The years since the launch of *Sputnik 1* in October 1957.

Space Exploration Initiative (SEI). Program proposed in July 1989 by President George Bush to return to the Moon and to go to Mars with people; terminated in 1993.

Space Shuttle. Informal name for the Space Transportation System, the reusable rocket system designed to transport people and machines to low Earth orbit.

Space Station. A permanent manned spacecraft in low Earth orbit; current version: the International Space Station.

spall. The rapid ejection of material from a planet by the passage of a shock wave through a planetary surface.

spectral data. Data dealing with the precise measurement of the color of planetary surfaces to determine composition.

spectroscopy. A measurement technique of determining energy intensity (reflected or emitted) as a function of wavelength.

spectrum. A plot of intensity as a function of wavelength.

spike. A rapid, unexpectedly sharp increase in a continuing process, for example, in the cratering rate.

spin axis. An imaginary line about which an object rotates.

Sputnik 1. The world's first artificial satellite, launched by the Soviet Union on October 4, 1957.

stable isotope. An isotope that is not radioactive, for example, oxygen isotopes.

stereo photographs. Two images that permit the three-dimensional nature of a planetary surface to be viewed.

stonewall. A hypothetical point beyond which the lunar radiometric "clock" cannot measure the ages of highland impact-melt breccias; supposedly occurs at 3.9 billion years.

Strategic Defense Initiative (SDI). Technology project of the 1930s designed to protect the United States from attack by missiles carrying nuclear warheads.

stratigraphy. The study of layered rock units; used to record the history of a planet.

strontium. A metal element; atomic number 38, atomic weight 87.6, symbol Sr.

subsatellite. A satellite ejected or launched from another satellite, versions of which were deployed by the *Apollo 15* and *16* missions in 1971 and 1972.

subsurface geology. The composition and structure of rock units below the surface of a planet.

sulfur. A light, nonmetallic element; atomic number 16, atomic weight 32.1, symbol S.

summit pit. A depression occurring at or near the top of a volcano.

supernova. An exploding star, producer of heavy elements; ejects particles that could trigger the formation of solar systems.

Surveyor. American space program of robotic precursor spacecraft that soft-landed on the Moon in 1966–68; documented the nature of the lunar surface in preparation for the Apollo missions.

synchronous rotation. Rotation in which the rate equals the period of revolution, so that the same face of a satellite always faces a planet; occurs for all regular moons in the solar system.

synodic period (month). Time between successive new moons; 29.5 days or 709 hours; equal to lunar day.

Synthesis Group. Presidential commission convened to examine program architectures for the Space Exploration Initiative in 1990–91.

syzygy. The alignment of the Sun, Earth, and the Moon, causing maximum tides.

tectonism. The process of deformation of planetary crusts and surfaces.

teleoperation. The remote control and operation of robotic machines by people.

telepresence. The collection of sensory input by a robotic machine and the transmittal of such sense data to a human controller at a remote location, resulting in the illusion of the controller being present at the remote site.

terminator. The line between day and night on the Moon.

terrae (TER-eye). The cratered highlands of the Moon (the Latin plural for "lands"; the singular is *terra* [TER-ah]).

terrestrial.   Of or like Earth.

terrestrial planet.   A rocky planet, similar in properties to Earth; group includes Mercury, Venus, Earth, Mars, and the Moon.

theory.   A hypothesis that has been continually and repeatedly tested and documented by experiment and observation to the point that it is nearly universally "accepted" by the scientific community.

thermal infrared.   The far infrared, where emission of thermal energy exceeds the reflectance of radiant energy; useful for remote sensing.

thorium.   A heavy metal element, radioactive, component of KREEP; atomic number 90, atomic weight 232, symbol Th.

tidal dissipation.   The frictional heat created by the tidal flexing of solid parts of planets; possibly an important source of heat early in lunar history.

tides.   The deformation of Earth and the Moon caused by their mutual gravitational attraction.

Titan 4.   The current U.S. heavy-lift booster, capable of lifting about 15 metric tons to low Earth orbit.

titanium.   A light metal element, common in lunar mare basalts; atomic number 22, atomic weight 47.9, symbol Ti.

topography.   "Hills and Holes": the irregularity of a planetary surface.

trace element.   A relatively scarce element that often contains critical clues to the origin and evolution of igneous rocks.

trailing edge.   The edge on the setting side of an object in the sky.

transit telescope.   A nonsteerable or fixed telescope that uses the rotation of the Moon to scan the sky.

traverse.   The following of a preplanned route over the surface of a planet, for example, to map and sample its geology.

troctolite.   A plutonic rock of the Mg-suite of the lunar highlands; made up of approximately equal amounts of the minerals plagioclase and olivine.

tungsten.   A metal element; atomic number 74, atomic weight 183.8, symbol W.

uranium.   A heavy metal element, radioactive; atomic number 92, atomic weight 238, symbol U.

velocity.   The property of speed and direction, expressed in many units—in this book, kilometers per second (km/sec).

vent.   A hole in a planet from which volcanic products (lava, ash) may be erupted.

vesicle.   A hole (frozen bubble) in a sample of lava rock; results from dissolved gas in the magma coming out of the solution.

viscosity.   The property that causes a fluid to resist flow or movement, with a higher viscosity resulting in slower rates of flow.

visible maria. The dark areas of the Moon, made up of lava flows erupted between 3.8 and about 1 billion years ago (some mare deposits may not be visible, having been ground up into the cratered terrain of the highlands).

volatile element. An element with a relatively low boiling temperature (e.g., hydrogen, sulfur).

volcanic rock. Rock formed by the cooling of liquid rock (lava) extruded onto a planetary surface.

volcanism. The planetary process of interior melting, the movement of the liquid rock through a planet, and the eruption of such liquids onto the surface.

wall terraces. A series of wall failures ("steps") caused by slumping in large impact craters.

wane. The apparent decrease in the lit area of the Moon over the course of the second half of a lunar day (two weeks).

wavelength. The distance between wave crests; for visible light, the same as color.

wax. The apparent increase in the lit area of the Moon over the course of the first half of a lunar day (two weeks).

weightlessness. "Zero-g"; experienced by free-falling (orbiting) objects in space.

wrinkle ridge. A tectonic feature created by compression in the maria of the Moon.

xenolith (ZEE-no-lith). Literally "stranger rock"; a fragment of rock from great depth carried to the surface as a clast in a lava flow or pyroclastic deposit.

X-rays. High-energy radiation occurring between ultraviolet and gamma-rays in a wavelength.

zenith. The point in the sky directly overhead, 90° from the horizon.

zero-g. The apparent absence of weight experienced by objects in free fall or in orbit; also called *microgravity*.

zinc. A light metal element; atomic number 30, atomic weight 65.4, symbol Zn.

A Lunatic's Library
An Annotated Bibliography of Important Books, Articles, and Videos about the Moon

The literature of the Moon is enormous, and the following list makes no claim to completeness. These are all books or sources that I believe are important in one way or another; all have their own bibliographies that will allow you to explore subjects in greater depth, if you so choose. Those books that are out of print can be found in most large public (and all university) libraries. Personal comments are entirely my own judgment.

Two Essential Books

Heiken, G. H., D. T. Vaniman, and B. M. French, eds. *The Lunar Sourcebook: A User's Guide to the Moon.* Cambridge: Cambridge University Press, 1991. 736 pp.
> The definitive reference book on the Moon. Team-written by over 30 active and former lunar scientists, this book is particularly thorough on lunar rocks and soils and, in this sense, nicely complements D. E. Wilhelms's book (see following entry). It is written for the layman but does not flinch on technical concepts.

Wilhelms, D. E. *The Geologic History of the Moon.* U.S. Geological Survey Professional Paper, no. 1348. Washington, D.C.: U.S. Government Printing Office, 1987. 302 pp.
> The historical geology of the Moon, written by one of the premier lunar geologists and historians of our science. Well written and beautifully illustrated, this book is a cogent summary of our understanding of the Moon from the stratigraphic perspective. Get it, read it, treasure it.

History Lesson

Chaikin, A. *A Man on the Moon.* New York: Viking Press, 1994. 670 pp.
> New book that deals with the Apollo program from the astronauts'

perspective. Very well done and interesting, it covers both science and operations.

Collins, M. *Carrying the Fire: An Astronaut's Journeys*. New York: Farrar, Straus & Giroux, 1974. 478 pp

The best book written by any astronaut, even if he does dislike geology! This is fascinating, funny, profound. Read this book and get a real feel for what going to the Moon was like.

Compton, W. D. *Where No Man Has Gone Before: A History of Apollo Lunar Exploration Missions*. NASA Special Publication, no. 4214. Washington, D.C.: U.S. Government Printing Office, 1989. 415 pp.

The "official" NASA history of Apollo lunar exploration. This book takes the stance that lunar flight was primarily a difficult engineering task, made even more difficult by whining scientists (a view with which I find myself in increasing agreement). It complements the scientific view by D. E. Wilhelms's *To a Rocky Moon*.

Hoyt, W. G. *Coon Mountain Controversies: Meteor Crater and the Development of Impact Theory*. Tucson: University of Arizona Press, 1987. 443 pp.

An exhaustive history of studies of Meteor Crater, Arizona, including much on the debate about the craters of the Moon. Very well written and enjoyable, it is highly recommended.

Lewis, R. S. *The Voyages of Apollo: The Exploration of the Moon*. New York: New York Times Book Company, 1974. 308 pp.

One of my favorite histories of the Apollo explorations, although it has been somewhat superseded by later accounts with more "perspective." It is still well worth reading.

McDougall, W. A. *The Heavens and the Earth: A Political History of the Space Age*. New York: Basic Books, 1985. 555 pp.

An exhaustive study of the politics of the space program and of government technology research in general. It is particularly heavy on the early, *Sputnik* days. The author is basically antagonistic toward government space programs, a view that is certainly different from that of most books in this field.

Murray, C., and C. B. Cox. *Apollo: The Race to the Moon*. New York: Simon and Schuster, 1989. 506 pp.

My favorite book about Apollo. This book wonderfully tells the engineering side of the story, including a nail-biting account of the near-disaster we almost had during the first landing on the Moon. It captures the excitement of the early days like no other book I know of.

Oberg, J. E. *Red Star in Orbit*. New York: Random House, 1981. 272 pp.

A well-written and fascinating book. Just what the heck *were* the

Soviets up to in space during the 1960s? What happened to the "Moon race," and why did the Russians never land a man on the Moon? These questions and others are answered here (to my satisfaction, anyway). I've been told that many Russians bought this book to learn about their own program!

O'Leary, B. *The Making of an Ex-Astronaut.* Boston: Houghton Mifflin, 1970. 243 pp.

The book that I love to hate. If you want to understand why Deke Slayton (1) didn't *want* any scientist-astronauts and (2) being forced to take them, didn't want to *fly* any scientist-astronauts, read this book, and it will all become clear. How O'Leary ever got selected as an astronaut is a mystery beyond all understanding.

Wilhelms, D. E. *To a Rocky Moon: A Geologist's History of Lunar Exploration.* Tucson: University of Arizona Press, 1993. 477 pp.

A wonderful book that is the clearest, most complete account of the history of lunar science in the space age. Weak on the early phases (which are well covered in Hoyt's book), it is unsurpassed for lunar science, starting with Ralph Baldwin and including geological mapping, astronaut training, and site selection. Lunar sample science awaits its Wilhelms to write its own history.

Wolfe, T. *The Right Stuff.* New York: Farrar, Straus & Giroux, 1979. 437 pp.

My favorite book about space, even though space is actually a marginal part of Wolfe's story. The quintessence of America in spirit and substance, this book is all the more startling in its contrast to the present space program and NASA. Read this and weep for your country.

Lunar Classics

Baldwin, R. B. *The Face of the Moon.* Chicago: University of Chicago Press, 1949. 239 pp.

The classic that got it all so right, so early. This book inspired Harold Urey in his interest in the Moon and greatly influenced many early lunar scientists.

Hartmann, W. K., R. J. Phillips, and G. J. Taylor, eds. *Origin of the Moon.* Houston: Lunar and Planetary Institute Press, 1986. 781 pp.

The proceedings of the Moon origin conference in Kona in 1984. Hardly a "classic" in the sense of some of these other books, this volume is the definitive statement of the "Big Whack" model for the origin of the Moon. The review papers by J. Wood, M. Drake, and L. Hood are particularly worthy; see also S. Brush's history of origin studies.

Mendell, W. W., ed. *Lunar Bases and Space Activities of the Twenty-First Century*. Houston: Lunar and Planetary Institute Press, 1985. 865 pp.
The proceedings of the 1984 Lunar Base Symposium in Washington, D.C. Great fun, this is a collection of wild fantasies about the advent of another Apollo program, come to save us all from the purgatory of space mediocrity. As one of my esteemed colleagues put it to me once, "We must have been smokin' dope!"

Mutch, T. A. *The Geology of the Moon: A Stratigraphic View*. Princeton: Princeton University Press, 1970. 324 pp.
An important book to me because it convinced me that if I wanted to study the Moon, I should change my major to geology. Much of this wonderfully written and illustrated account of the stratigraphy (layered rocks) of the Moon is still current (although some outdated concepts are here, like highland volcanic domes).

Schultz, P. H. *Moon Morphology: Interpretations Based on Lunar Orbiter Photography*. Austin: University of Texas Press, 1976. 626 pp.
A massive compilation of Lunar Orbiter images of just about every imaginable lunar feature, classified by type of landform. Although the images are beautifully reproduced, the book's value is somewhat diminished by some strange interpretations and the fact that it is limited to only Lunar Orbiter and does not include Apollo pictures (for that, see Masursky, Colton, and El-Baz, *Apollo over the Moon*).

Taylor, S. R. *Lunar Science: A Post-Apollo View*. New York: Pergamon Press, 1975. 372 pp.
Thorough, well-documented summary of the results of lunar sample science up to about 1975. Read this to see how we perceived the Moon immediately following Apollo. If I had been writing this book 15 years ago, Taylor's and Mutch's books would be my "two essentials of lunar science."

Urey, H. C. *The Planets: Their Origin and Development*. New Haven: Yale University Press, 1952. 245 pp.
Of historical interest only. Chapter 2 is on the Moon. Urey believed in going back to first principles of physics and in deducing everything about the Moon from a few hard facts. He was also famous for believing that geologists were "second-rate scientists." Urey was wrong in just about all of his beliefs about the Moon, and I'm happy to report that he lived to see the complete vindication of the pseudoscientific, geological point of view from the Apollo data.

Readable and Reliable Popular Accounts of Lunar Science and Exploration

Cooper, H. F. S. *Apollo on the Moon*. New York: Dial Press, 1969. 144 pp.
    Instant book from the *New Yorker* reporter who covered Apollo. That
    being so, the book is a well-written, brief summary of the first lunar
    landing.

———. *Moon Rocks*. New York: Dial Press, 1970. 197 pp.
    A marvelous and amusing work that nicely captures both the pompos-
    ity and the occasional brilliance of the "scientific community." While
    writing *Apollo on the Moon*, Cooper got interested in the work of the
    Preliminary Examination Team (PET), the group of scientists who
    were the first to examine the *Apollo 11* moon rocks after the mission.
    He wrote this book as a result.

Cortwright, E. M., ed. *Apollo Expeditions to the Moon*. NASA Special
Publication, no. 350. Washington, D.C.: U.S. Government Printing Office, 1975. 313 pp.
    A collection of essays on all aspects of the Apollo program, from
    booster rockets to lunar science, written by participants. Beautifully
    illustrated with many color photographs, this book should have been
    reprinted by NASA to celebrate the 25th anniversary of *Apollo 11*.

French, B. M. *The Moon Book*. New York: Penguin Press, 1977. 287 pp.
    Very nice account of what the lunar samples tell us, written by a
    sample scientist. This book is similar in concept and scope to the one
    you are reading, only French's book was written 20 years ago; some
    things have changed a lot . . . (others haven't—see his words on the
    Lunar Polar Orbiter mission on p. 265!).

Lewis, J. S., and R. A. Lewis. *Space Resources: Breaking the Bonds of
Earth*. New York: Columbia University Press, 1987. 418 pp.
    A strange book that I cannot make up my mind about. Basically a
    paean to mining asteroids, it correctly analyzes many of the problems
    with our current space program, but some perspectives seem off-
    center to me. (For example, the Lewises argue that huge space pro-
    grams like the Shuttle eat up small, wonderful unmanned programs,
    but the best unmanned explorations we've ever had [Mariner, Viking,
    Voyager] all derived from the exploratory momentum of the Apollo
    project.) However, the authors' contentions about a lack of direction
    in the space program and their ideas on how to fix this problem seem
    pretty good. Read the book and decide for yourself.

Masursky, H., G. W. Colton, and F. El-Baz, eds. *Apollo over the Moon: A
View from Orbit*. NASA Special Publication, no. 362. Washington, D.C.:
U.S. Government Printing Office, 1978. 255 pp.

A beautifully illustrated collection of the very best photographs taken from lunar orbit during the Apollo missions, each one presented with a geologically oriented caption by a relevant expert. This is another wonderful NASA book that is out of print but well worth the search and perusal.

Wood, J. A. *The Solar System.* Englewood Cliffs, N.J.: Prentice-Hall, 1979. 196 pp.

Nice summary of various geological processes in the solar system, including many of the results from study of the lunar samples. Data are placed into a broader scientific context than is usual for books like this.

Book and Magazine Articles

Burns, J. O., N. Duric, G. J. Taylor, and S. W. Johnson. "Observatories on the Moon." *Scientific American* 262, no. 3 (March 1990): 42–49.

Details the advantages and potential problems of conducting astronomy on the Moon.

Burt, D. M. "Mining the Moon." *American Scientist* 77, no. 6 (November-December 1989): 574–79.

Concise summary of what the Moon has to offer by way of materials and how we might extract what we need from it.

Editors of Time-Life. "Earth's Companion." In *Moons and Rings*, 6–49. A book in the "Voyage through the Universe" series. Alexandria, Va.: Time-Life Books, 1991.

Good, general treatment of the Moon, including some very nice artwork from one of my favorite artists of the Moon, Don Davis. (See also the Time-Life book *Solar System* in the "Planet Earth" series; Davis has a series of magnificent lunar paintings in a spread called "Life Cycle of the Solar System.")

Ryder, G. "Apollo's Gift: The Moon." *Astronomy* 22, no. 7 (July 1994): 40–45.

Nice summary of the evolution of the Moon, inferred from study of the lunar samples. This article is well illustrated with cross sections (in color) showing the interior of the Moon through time.

Spudis, P. D. "An Argument for Human Exploration of the Moon and Mars." *American Scientist* 80, no. 3 (May-June 1992): 269–77.

My attempt to identify the unique capabilities of people as scientific explorers in space. This failed to help the Space Exploration Initiative to any significant degree; the scientific community doesn't really want to confront this issue.

———. "The Moon." In *The New Solar System*, 3d ed., edited by J. K. Beatty and A. Chaikin, 41–52. Cambridge: Cambridge University Press; Cambridge, Mass.: Sky Publishing, 1990.

> A summary written before the Clementine mission, so it is somewhat out of date but still largely contemporary. B. French wrote the chapter on the Moon in the first two editions of this readable, colorfully illustrated book.

Taylor, G. J. "The Scientific Legacy of Apollo." *Scientific American* 271, no. 1 (July 1994): 26–33.

> A well-written summary of our current understanding of the Moon, by my colleague and sometime collaborator, an active lunar scientist.

Taylor, S. R. "The Origin of the Moon." *American Scientist* 75, no. 5 (September–October 1987): 469–77.

> A clear and succinct statement of the "Big Whack" hypothesis. Taylor is a believer; I'm not so sure.

Wilhelms, D. E. "Moon." In *The Geology of the Terrestrial Planets*, edited by M. H. Carr, 107–205. NASA Special Publication, no. 469. Washington, D.C.: U.S. Government Printing Office, 1984.

> Actually a brief "abstract" version of Wilhelms's book that started this bibliographic review. It is well worth reading, even though a few concepts are out of date. The color "paleogeological" maps of the Moon at the end are very nicely done and allow you to visualize how the Moon's surface has changed over time.

Two Lunar Atlases and a Map

Bowker, D. E., and J. K. Hughes. *Lunar Orbiter Photographic Atlas of the Moon.* NASA Special Publication, no. 206. Washington, D.C.: U.S. Government Printing Office, 1971. 41 pp., 675 plates.

> The definitive collection of Lunar Orbiter pictures, showing almost the entire lunar surface, near and far sides. Its value is somewhat hampered by the relatively poor reproduction of some of the photographs. Out of print, it might be found in large libraries, particularly in those having U.S. government document sections.

Rükl, Antonín. *Hamlyn Atlas of the Moon*, edited by T. W. Rackham. London: Paul Hamlyn Publishing, 1990. 224 pp.

> Excellent atlas of the near side of the Moon. Particularly useful for amateur astronomers and observers, Rükl's careful drawings can be enjoyed by all students of the Moon. Each map in the atlas gives a brief entry on the people for whom craters were named. A major drawback is that there is no comparable version for the far side, but

with new data from the Clementine mission, maybe there will be soon. This atlas is available directly by mail order from *Astronomy* magazine.

National Geographic Society. *The Earth's Moon*, 2d ed. Washington, D.C.: Cartographic Division, National Geographic Society, 1976.
   The best map of the Moon, showing both near and far sides (with complete and current feature names) on a single sheet at a scale of 1:10,000,000. The margins are filled with fascinating facts and drawings about the Moon and an index of named formations. (A note of caution to those who get their maps from used-book dealers: There are two versions of this map in existence. The first edition, published in 1969 [to coincide with the Apollo missions], does not have names for far-side features. Be sure to get the current edition; if you can't find it used, the Society sells it from its catalog.)

Moon Lore and Mythology

Brueton, D. *Many Moons*. New York: Prentice-Hall Press, 1991. 256 pp.
Guiley, R. E. *Moonscape*. New York: Prentice-Hall Press, 1991. 192 pp.
   A poser for you: Why would the same publisher come out with two books of nearly identical scope and subject matter in the same year? Their Moon must have been in the Seventh House. Anyway, I actually find these books quite enchanting. They are both wonderful collections of various lunar myths, creation stories, magic, lycanthropy, and various other New Age and multicultural minutiae. The late Jim Irwin (*Apollo 15* LM pilot) wrote the foreword to *Many Moons*—and is inexplicably identified on the dust jacket as "of the *Apollo 11* Mission"!

Movies and Videos

Documentaries and History

PBS. *Spaceflight*. Pacific Arts Video, 1985. 4 parts, 240 min.
   A reasonably good series that tells the story of the race to the Moon. The whole series is worth watching, but the best episode is part 3, "One Giant Leap," in which the moon landing of *Apollo 11* is recounted.
Reinart, A. *For All Mankind*. National Geographic Video, 1989. 79 min.
   A documentary made up of footage from all of the Apollo missions, artistically combined into a single continuous narrative on how we explored the Moon. Glorious!

TBS Productions. *Moonshot.* Turner Home Video, 1994. 2 parts, 240 min.

Latest in the "story of the Moon race" sweepstakes. This is narrated by actor Barry Corbin, who portrays Deke Slayton. The many personal reminiscences by astronauts are fascinating. The first half is much better than the second. There are many interesting segments, including the best account of the cliff-hanging *Apollo 11* first lunar landing.

WQED. *To the Moon and Beyond.* Episode 4 in *Space Age.* Public Media Video, 1992. 60 min.

A well-made, cogent work on a return to the Moon. The National Academy of Sciences offered us the series *Space Age* as its contribution to International Space Year. The effort was largely a bust (as was the ISY), except for this episode, which summarizes the major themes outlined in Chapters 9 and 10 of this book.

Three Lunar Feature-Film Classics

Let us finally pay homage to the power of imagination. In my opinion, only three movies dealing with lunar spaceflight are worth anything.

Apollo 13. Brian Grazer, Producer. Universal Pictures, 1995. 130 min.

The newest entry into the lunar film sweepstakes (having come out while I was writing this book). I may be premature in adding this to my list now, but my initial impression is that this film beautifully captures the spirit and substance of Apollo spaceflight. Space scenes are realistic (many were filmed in the real "microgravity" of the NASA KC-135 aircraft) and gripping. As usual with films about the Moon (except for *Destination Moon*), many liberties are taken with lunar geography (e.g., while flying over Tsiolkovsky crater, on the far side, the crew says that they "can look up towards Mare Imbrium," which is of course on the opposite hemisphere).

Destination Moon. George Pal, Producer. Sinister Cinema Video, 1950. 91 min.

Based on a short story by Robert Heinlein. Back in the paleolithic dawn of the space age, this film tried to "educate" the public about things to come. Beautifully done, it includes wonderful space art by Chesley Bonstell. Listen to the description of the Moon by the spaceship crew and compare it to the uncannily similar words of Neil Armstrong and Buzz Aldrin, only 20 years (but an emotional lifetime) later.

2001: A Space Odyssey. Stanley Kubrick, Producer. MGM Video, 1968. 139 min.

The ultimate space movie—philosophical, intellectual, emotional, profound. This film takes great pride in getting every technical detail right, even down to the subtleties of weightlessness and artificial gravity. So how come the Moon changes its phase forward, backward, and in eight-day leaps during the voyage between the space station and Clavius Base (a lousy place for a lunar outpost, by the way)? Also, during the scene at Tycho, having Earth appear so low on the horizon is wrong (Tycho is at 43° S latitude, so Earth would appear halfway between the horizon and directly overhead). Still, there is nothing else like it for the real "feel" of spaceflight.

Index

absolute zero, 187
absorption band, 174–75
accretion, 95, 146, 153
aerobrakes, 205
ages: absolute, 50–52, 64–65, 67, 98, 141, 144, 229; crystallization, 50; relative, 17, 20, 48, 50–52
agglutinate, 89–91, 92, 132
Albritton, Claude, 17
Alphonsus crater, 58–59, 112
ALSEP (Apollo Lunar Surface Experiment Package), 66, 71, 223
angular momentum, 161
annealing, 135
anorthosite, 60, 66, 147–49, 153; ages of, 148, 149
Antarctica, 239
Anti-Ballistic Missile Treaty, 183
Apennine Bench Formation, 150
Apennine Mountains, 48, 49, 72, 143
apogee, 10
Apollo 1 fire, 180
Apollo 8, 62–63
Apollo 10, 63
Apollo 11, 63–66, 103, 105, 96, 114, 147
Apollo 12, 60, 66–68, 96, 106, 150
Apollo 13, 68–69
Apollo 14, 69–70, 123, 138, 143, 150, 227
Apollo 15, 72–73, 87, 106, 103, 118–19, 139, 143, 150, 152, 222
Apollo 16, 34, 73–76, 107, 139, 142, 144, 148, 227
Apollo 17, 50, 76–77, 106, 108, 111, 143, 144, 152, 170, 217, 222, 235–36
Apollo program, 53–56, 211; extended ("J-missions"), 70; mode decision, 54; samples from, 79; uniqueness of, 250
Archimedes crater, 48, 49
Arecibo Observatory, 200
Aristarchus, 2
artificial gravity, 210–11
ash, 42, 73, 77, 103–13, 130, 169, 206
assimilation, 153
asteroid, Geographos, 186
astroblemes, 17
astronomy, 196–202; problems with lunar, 202
atom bomb, 251
Augustine, Norman, 182
Augustine Committee, 182

Baldwin, Ralph, 19–20, 102, 190
Barringer, Daniel, 17
basalts, 20, 39, 60, 64–65, 67, 70, 77, 81, 102–3; amount in crust, 108; ancient, 107, 123, 125; high-aluminum, 127; high-titanium, 106, 125; low-titanium, 106, 125, 127
basins, 20, 29, 194; ejecta, 32, 144; filling, 129–30; grooved terrain, 34; impact melt, 36, 50; loading, 122–23; rings, 36, 37; secondaries, 34
bedrock, 84
Beer, Wilhelm, 5
bombardment, 83, 84, 191; early, 96, 146–47; periodic, 98
Boon, John, 17
breccia, 64, 68, 69, 73, 74, 94, 131–35; clasts in, 132; fragmental, 74, 134–35; granulitic, 135, 152; matrix, 132; types of, 134

Du, 132